INTERNATIONAL CENTRE FOR MECHANICAL SCIENCES

COURSES AND LECTURES - No. 211

MULTICRITERIA DECISION MAKING

EDITED BY

G. LEITMANN

UNIVERSITY OF CALIFORNIA, BERKELEY

A. MARZOLLO

UNIVERSITY OF TRIESTE

SPRINGER-VERLAG WIEN GMBH

Originally published by Springer-Verlag Wien-New York in 1975

ISBN 978-3-211-81340-9 ISBN 978-3-7091-2438-3 (eBook)
DOI 10.1007/978-3-7091-2438-3

LIST OF CONTRIBUTORS

A. Blaquière	Laboratoire d'Automatique Théorique, Université de Paris VI, Paris, France
G. Castellani	Department of Mathematics, Ca' Foscari University, Venice, Italy
G. Leitmann	Mechanical Engineering Department, University of California at Berkeley, California, USA
Y. Medanic	Mihailo Pupin Institute, Belgrade, Yugoslavia
A. Marzollo	Electrical Engineering Department, University of Trieste and International Centre for Mechanical Sciences, Udine, Italy
W. Stadler	Mechanical Engineering Department, University of California at Berkeley, California, USA
W. Ukovich	Electrical Engineering Department, University of Trieste, Italy
M. Volpato	Department of Mathematics, Ca' Foscari University, Venice, Italy
P.L. Yu	Graduate School of Business, University of Texas, Austin, Texas, USA.

CONTENTS

PREFACE

A considerable amount of research has been devoted recently to Multicriteria Decision Making, stimulated by the vast number of real problems, for example in industrial, urban and agricultural economics, in the social sciences, and in the design of complex engineering systems, where many decision makers are present or many, possibly conflicting objectives should be taken into account in order to reach some form of optimality.

A rough division into two classes may be made among the approaches to Multicriteria Decision Making problems. The first one deals mainly with the empirical determination of preference structures in some specific problems, and seeks methods for their meaningful aggregation in order to arrive, often by ad hoc and iterative procedures, at practically reasonable solutions. The lines of thought which are followed and the used methods may be looked upon as modern developments of operations research.

The second one, predominantly treated by researches whose background is often rooted in systems and control theory, or in mathematical programming and in its applications, seems more directed toward general and rigorous formulations in order to reduce Multicriteria Decision Problems conceptually to clearly defined classes of optimization problems for which definite solutions algorithms are sought.

The contributions to the present volume follow mainly the latter line of thought, although references and comparisons are made to other, sometimes non-empirical methods, for example by P.L. Yu and, in general, algorithmic solutions are proposed to specific problems as a result of the conceptual methods used. The reader is introduced to the large class of multicriteria, multiagent problems which may be treated in the framework of game theory, both for static and dynamical systems, in the first two contributions by G. Leitman and A. Blaquière. The formulation of the latter is so general as to encompass as specific cases the great majority of multi-objective, multiplayer problems that one may think of in cooperative, non-cooperative or mixed situations.

J. Medanic gives an exhaustive solution, both in deterministic and stochastic cases, to the optimal regulator problem with vector valued quadratic performance, and applies the developed concepts to a multiplant cooperative control problem.

The paper by W. Stadler imbeds both the static and dynamical vector optimization problem in the framework of preference optimality by borrowing techniques which have been developed mainly by mathematical economists, and so is able to give interesting sufficient and necessary conditions for optimality; vector optimization concepts are then applied to the design of minimally disturbing measuring devices for optimally controlled systems, and for optimal structural design in mechanics. A complete treatment of Domination Structures and Non-Dominated Solutions with an example of application to stock market behaviour is given by P.L. Yu.

A Marzollo and W. Ukovich discuss some basic principles underlying the concepts of Vector Optimality and then give precise conditions, using the weakest hypotheses on the involved functions, for the characterization of "weakly" "ordinarily" and "strictly" Non Dominated Decisions, in the global, local and "differential" version.

Specific economic relevance is stressed in the contribution of M. Volpato, who deals with the optimal choice for the relative amounts of a specific good to be produced in various countries of a community in order to give the maximal community yield from the residual resources. Unlike the classical theory on the subject, non-linear conversion prices from one product to another are also considered. The problem may be solved explicitely for a rather general class of functions by using non-linear and dynamic programming techniques which are developed in the following contribution by G. Castellani. It is shown that in this framework the optimal production policy for the community is also economically optimal for each individual country.

This sketch of the contents of this volume is by no means exhaustive of the various phylosophical approaches to Multicriteria Optimization contained therein, nor of the techniques which, as a consequence, are suggested for the solution of many varied problems. We express our hope that the volume as a whole will stimulate the reader to giving further thought to the conceptual and mathematical challenges offered by the present extension of optimality theory and provide him with some useful methods for solving problems in which different points of view are to be considered, or, to recall the title of this volume, to solve "Multicriteria Decision" problems.

George Leitmann Angelo Marzollo

COOPERATIVE AND NON-COOPERATIVE DIFFERENTIAL GAMES

G. Leitmann

Department of Mechanical Engineering

University of California, Berkeley

ABSTRACT. Many player differential games are discussed for a cooperative mood of play in the sense of Pareto, and for a non-cooperative one, in the sense of Nash. In the cooperative case, the results are equally applicable to the situation of a single decision-maker with multi-criteria. Necessary as well as sufficient conditions for optimal play are considered. Some examples are presented to illustrate the theory.

1. INTRODUCTION

1.1 Problem Statement

We consider games which involve a number of players. The rules of the game assign to each player a cost function of all the player's decisions and the sets from which these decisions can be selected.

Let there be N players. Let $J_i(\cdot)$ and D_i be the cost function and decision set, respectively, for player i. Then.

$$J_i(\cdot) : D \to R^1 \qquad i = 1,2,\ldots,N \tag{1.1}$$

where $D \subseteq \prod_{i=1}^{N} D_i$.

Loosely speaking, each player wishes to attain the smallest possible cost to himself. Thus, if there is a $d^u \in D$ such that for all $i \in \{1,2,\ldots,N\}$

$$J_i(d^u) \leq J_i(d) \qquad \forall d \in D \tag{1.2}$$

then d^u is certainly a desirable decision N-tuple (joint decision). Unfortunately, such a <u>utopian</u> (absolutely cooperative) decision rarely exists (e.g., Refs. 1 - 3) and the players are faced with a dilemma: What mood of play should they adopt, that is, how should an "optimal" decision be defined?

Here we consider only two moods of play, one cooperative and the other non-cooperative, in the sense of Pareto[4] and Nash,[5] respectively.

1.2 Cooperative Play

If the players decide to "cooperate" in making their individual decisions, they can do so by adopting a joint decision as suggested by Pareto. There is more than one way of defining Pareto-optimality. One of these is given by

Definition 1.1. A decision N-tuple $d^o \in D$, $D \subseteq \prod_{i=1}^{N} D_i$, is Pareto-optimal iff for every $d \in D$ either

$$J_i(d) = J_i(d^o) \qquad \forall i \in \{1,2,\dots,N\}$$

or there is at least one $i \in \{1,2,\dots,N\}$ such that

$$J_i(d) > J_i(d^o) \quad .$$

In this definition of Pareto-optimality cooperation is embodied in a statement such as "I am willing to forego a gain (a decrease in my cost) if it is to be at the expense of one of the other players (an increase in his cost)."

Alternatively, we can state the equivalent

Definition 1.2. A decision N-tuple $d^o \in D$, $D \subseteq \prod_{i=1}^{N} D_i$, is Pareto-optimal iff for all $d \in D$

$$J_i(d) \leq J_i(d^o) \qquad \forall i \in \{1,2,\dots,N\}$$

implies

$$J_i(d) = J_i(d^o) \qquad \forall i \in \{1,2,\dots,N\} \quad .$$

This way of defining Pareto-optimality leads to a statement such as "If a joint decision is not Pareto-optimal, then there is another decision that results in the decrease of at least one cost without increasing any of the others."

In Figure 1.1 there is an illustration of a simple example of Pareto-optimality. Contours of constant cost for each player are plotted in joint decision space, $\pi_{i=1}^{N} D_i$; the Pareto-optimal decision N-tuples are points of tangency of cost contours.

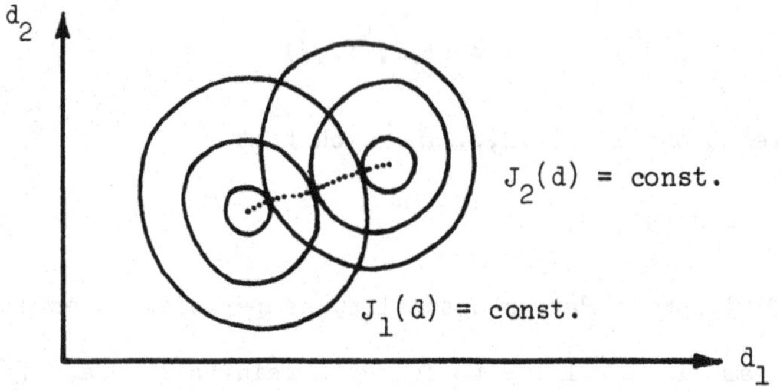

Fig. 1.1, Pareto-optimality

There is yet another equivalent definition of Pareto-optimality.

Let

$$J(d) = \{J_1(d), J_2(d), \ldots, J_N(d)\}$$

and

$$Y = \{J(d) : d \in D\}$$

and

$$\Lambda^{\leqq}(J(d)) = \{y \in Y : y_i - J_i(d) \leqq 0\} - J(d)$$

That is, $\Lambda^{\leqq}(J(d))$ is the non-positive orthant with vertex at $J(d)$ minus its vertex.

Then we have

<u>Definition 1.3</u>. A decision N-tuple $d^o \in D$, $D \subseteq \pi_{i=1}^{N} D_i$, is Pareto-

optimal iff $\Lambda^{\leq}(J(d^o)) \cap Y = \phi$.

This definition is illustrated in Figure 1.2. It is readily seen that cost N-tuples, $J(d^o)$, corresponding to Pareto-optimal decisions, d^o, belong to the boundary of the set of feasible costs, Y.

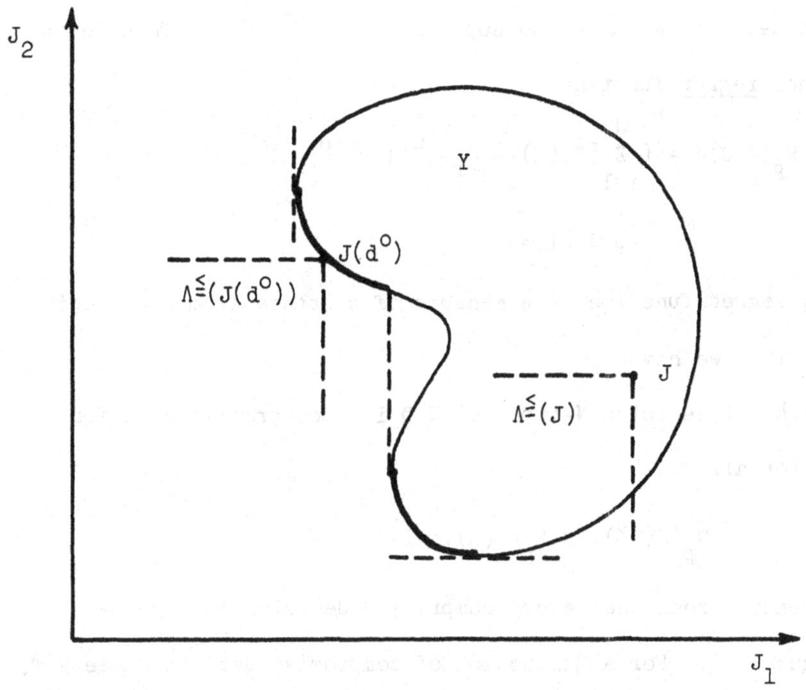

Fig. 1.2, Feasible Cost Set

If cooperation is defined in the sense of Pareto then, given a decision N-tuple $d \in D$ with $\Lambda^{\leq}(J(d)) \cap Y \neq \phi$, one can conclude that there is a $d' \in D$ such that $J_i(d') < J_i(d)$ for some $i \in \{1,2,\ldots,N\}$ where $J(d') \in \Lambda^{\leq}(J(d))$. That is, d' is "preferred" to d. However, nothing concerning preference can be said for decision N-tuples d'such that $J(d') \notin \Lambda^{\leqq}(J(d))$. Thus, only one orthant of cost space is utilized for comparing decision N-tuples. To overcome this restriction Yu has introduced the notion of "domination structures;" see the chapter Domination Structures and Nondominated Solutions by P. L. Yu.

Another appealing way of defining optimality, related to Pareto-optimality, involves the introduction of a regret function. Let $d^u \in D$ be a utopian decision N-tuple and suppose that $J(d^u) \notin Y$. Then define the p-th order regret function

$$R_p(J(d)) = \{\sum_{i=1}^{N} [J_i(d) - J_i(d^u)]^p\}^{1/p}$$

$$p \in [1,\infty]$$

That is, the regret function is a measure of distance from the utopia cost $J(d^u)$. Then we have

Definition 1.4. A decision N-tuple $d^c \in D$ is a compromise decision of order p iff for all $d \in D$

$$R_p(J(d^c)) \leqq R_p(J(d)) \quad .$$

It is readily seen that every compromise decision is Pareto-optimal, Figure 1.3. For a discussion of compromise decisions see Ref. 3.

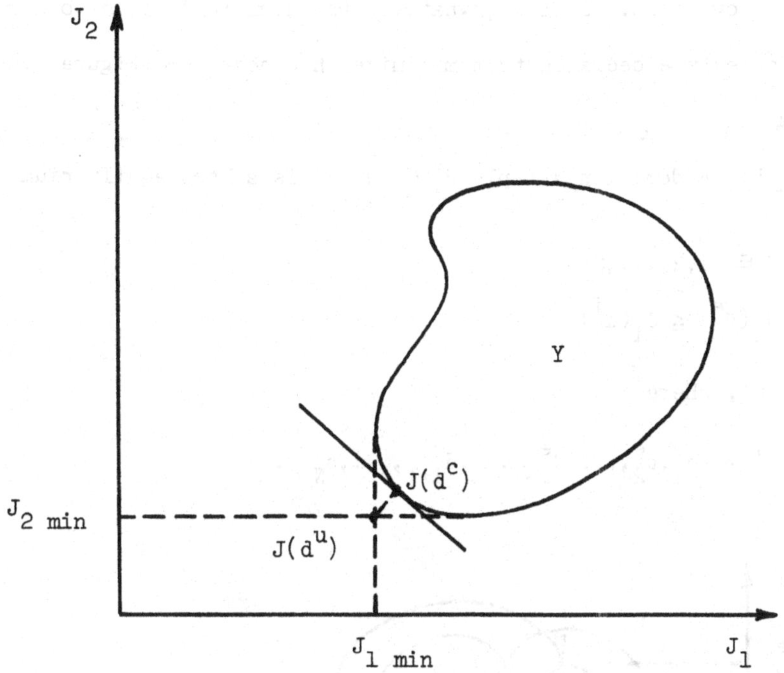

Fig. 1.3, Compromise Decision for p = 2.

Finally, it should be noted that the aforegoing discussion applies equally well to the case of a single decision-maker with more than one cost function.

1.3 Non-Cooperative Play

If the players do not cooperate, but rather if each player strives to minimize his own cost regardless of the consequence to the other players, then each player is faced with a problem: In selecting his "best" decision, what should he assume about the other players' decisions?

In Nash's definition of optimality, every player assumes that each of the other players makes his decision only with a view toward minimizing his own cost. That is, whatever the decisions of the other players, he selects a decision that minimizes his cost; see Figure 1.4. Thus one has

Definition 1.5. A dedision N-tuple $d^e \in \prod_{i=1}^{N} D_i$ is a Nash equilibrium iff for all $i \in \{1,2,\ldots,N\}$

$$J_i(d^e) \leq J_i(d^i)$$

for all $d_i \in D_i$, where

$$d^i = \{d_1^e, d_2^e, \ldots, d_{i-1}^e, d_i, d_{i+1}^e, \ldots, d_N^e\} .$$

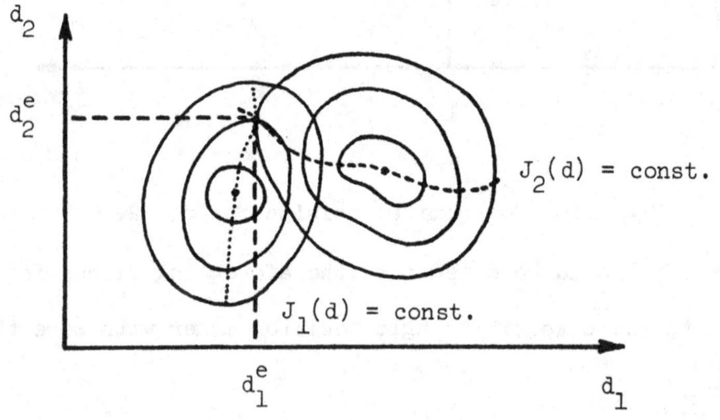

Fig. 1.4, Nash Equilibrium

Thus, a Nash equilibrium is characterized by the fact that no one player can improve his position (decrease his cost) by adopting another decision, provided the remaining players stick with their equilibrium decisions. Of course, that does not imply that by cooperating the players might not be able to decrease some or even all of the costs without increasing any.

An important class of Nash equilibrium games is that of two-person zero-sum games. In these games one player loses what the other player gains; that is,

$$J_1(d) = - J_2(d) \triangleq J(d) \tag{1.3}$$

Hence, in terms of J(d), player 1 is the minimizer and player 2 is the maximizer. We have

Definition 1.6. Decision couple $d^e \in D_1 \times D_2$ is a saddlepoint iff

$$J(d_1^e, d_2) \leq J(d_1^e, d_2^e) \leq J(d_1, d_2^e)$$

for all $d_1 \in D_1$, $d_2 \in D_2$.

Two-person zero-sum games played on crossproduct decision sets, $D_1 \times D_2$, have a number of desirable properties not shared by nonzero-sum games. Among these are, e.g., see Ref. 6:

(i) If d^e and $\bar{d}^e$ are saddlepoints, then

$$J(d^e) = J(\bar{d}^e) \triangleq J^e$$

where J^e is termed the Value of the game.

(ii) If d^e and $\bar{d}^e$ are saddlepoints, then

$$J(d_1^e, \bar{d}_2^e) = J(\bar{d}_1^e, d_2^e) \quad .$$

(iii) $$\bar{J} \geq \underline{J}$$

where

$$\bar{J} = \inf_{D_1} \sup_{D_2} J(d_1, d_2)$$

$$\underline{J} = \sup_{D_2} \inf_{D_1} J(d_1,d_2)$$

If a saddlepoint, d^e, exists

$$\text{(iv)} \qquad \bar{J} = \underline{J} = J^e$$

and

$$\text{(v)} \qquad J(d_1^e,d_2) \leqq \bar{J} \quad \forall d_2 \in D_2$$

$$J(d_1,d_2^e) \geqq \underline{J} \quad \forall d_1 \in D_1$$

Thus, according to i), all saddlepoints are equivalent in terms of cost. According to ii), each player can choose any saddlepoint decision, in case of non-uniqueness, and still assure the Value of the game. Finally, in view of (v), by adopting a saddlepoint decision, a player can assure a cost, regardless of his opponents' decision, that is at least as favorable as the best he can guarantee.

2. DYNAMICAL SYSTEM

2.1 State Equation

We are concerned with a dynamical system defined by its state, a point $x \in R^n$, which changes in a prescribed manner with the passing of time $t \in (-\infty,\infty)$; of course, any time-like variable can serve in place of time. The evolution of the state is controlled by N players.

Given an initial state x_0 at time t_0, let $\tau = t - t_0$. Furthermore, let the n-th component of x be t itself; that is, $x_n \equiv t$.

Consider functions

$$u^k(\cdot) : [0,\tau_1] \to R^{d_k} \quad , \quad k = 1,2,\ldots,N$$

and C^1 functions

$$f(\cdot) : R^{n+d_1+\ldots+d_N} \rightarrow R^n$$

and let dot denote differentiation with respect to τ. The evolution of the state is described by an absolutely continuous function

$$x(\cdot) : [0,\tau_1] \rightarrow R^n \text{ , } x(0) = x_0$$

satisfying a given state equation

$$\dot{x}(\tau) = f(x(\tau),u^1(\tau),u^2(\tau),\ldots,u^N(\tau)) \qquad (2.1)$$

2.2 Controls and Strategies

The players influence the evolution of the state through their choices of the $u^k(\tau)$ for almost all $\tau \in [0,\tau_1]$. We consider two ways of making these choices: The players use either (relative) time τ or state $x(\tau)$ as the information on which to base their choices.

In the former case, applicable in cooperative games, each player selects an admissible open-loop control; that is, player k chooses a Lebesgue measurable, bounded function

$$u^k(\cdot) : [0,\tau_1] \rightarrow R^{d_k}$$

In the latter case, applicable in non-cooperative games, each player selects an admissible closed-loop or feedback strategy; that is, player k chooses a Borel measurable, bounded function

$$p^k(\cdot) : R^n \rightarrow R^{d_k}$$

so that, in (2.1),

$$u^k(\tau) = p^k(x(\tau)) \qquad (2.2)^+$$

+Since $p^k(\cdot)$ is Borel measurable and $x(\cdot)$ is absolutely continuous, $u^k(\cdot)$ defined by (2.2) is Borel and hence Lebesgue measurable.

In addition, admissible controls and strategies may be subject to constraints. For instance, given $U^k \subseteq R^{d_k}$, it may be required that

$$u^k(\tau) \in U^k \text{ a.e. } \tau \in [0,\tau_1]$$

and

$$p^k(x) \in U^k \text{ a.e. } x \in R^n .$$

2.3 Playability*

Whether playing cooperatively or not, we suppose here that all players desire to steer the state from a given initial state x_0 to a state on a prescribed target set $\theta \subset R^n$. We have

Definition 2.1. A control N-tuple

$$u(\cdot) = \{u^1(\cdot), u^2(\cdot), \ldots, u^N(\cdot)\} : [0,\tau_1] \to R^{d_1+d_2+\ldots+d_N}$$

is playable at x_0 iff it is admissible and generates a solution $x(\cdot)$ such that $x(0) = x_0$ and $x(\tau_1) \in \theta$. A strategy N-tuple

$$p(\cdot) = \{p^1(\cdot), p^2(\cdot), \ldots, p^N(\cdot)\} : R^n \to R^{d_1+d_2+\ldots+d_N}$$

is playable at x_0 iff it generates at least one solution $x(\cdot)$ such that $x(0) = x_0$, $x(\cdot) \notin \theta$ for $\tau \in [0,\tau_1]$, and $x(\tau_1) \in \theta$. A corresponding triple $\{x_0, p(\cdot), x(\cdot)\}$ is termed a terminating play.

2.4 Performance Index

Associated with each player there is a performance index (cost function). For a cooperative game in which the players use controls,

---

*For games, such as pursuit-evasion ones, in which one player desires termination while the other one does not, see "games of kind" in Ref. 13 and "qualitative games" in Ref. 15.

the costs depend on initial state x_0 and admissible control N-tuple $u(\cdot)$, the corresponding solution $x(\cdot)$ being unique, whereas in the non-cooperative case of the players employing strategies, the costs depend on initial state x_0, admissible strategy N-tuple $p(\cdot)$ and a corresponding solution $x(\cdot)$. We take the cost for player i to be either

$$\begin{matrix} V_i(x_0,u(\cdot)) \\ \text{or} \\ V_i(x_0,p(\cdot),x(\cdot)) \end{matrix} = \int_0^{\tau_1} f_0^i(x(\tau),u^1(\tau),\ldots,u^N(\tau))d\tau \qquad (2.3)$$

where $f_0^i(\cdot) : R^{n+d_1+\ldots+d_N} \to R^1$ is C^1. Of course, in the latter case $u^k(\cdot)$ is given by (2.2) on some bounded interval $[0,\tau_1]$.

3. COOPERATIVE DIFFERENTIAL GAMES

3.1 Pareto-Optimality

For a dynamical system, Definition 1.1, for instance, becomes

Definition 3.1. A control N-tuple $u^0(\cdot)$, playable at x_0, is Pareto-optimal iff for every control N-tuple $u(\cdot)$, playable at x_0, either

$$V_i(x_0,u(\cdot)) = V_i(x_0,u^0(\cdot)) \quad \forall i \in \{1,2,\ldots,N\}$$

or there is at least one $i \in \{1,2,\ldots,N\}$ such that

$$V_i(x_0,u(\cdot)) > V_i(x_0,u^0(\cdot)) .$$

3.2 Necessary Conditions

By means of the following, readily established lemma and corollary (Ref. 7) one can reduce necessary conditions for Pareto-optimal control to ones for an optimal control problem with isoperimetric (integral) constraints (e.g., Refs. 8 - 9).

Lemma 3.1. If $u^0(\cdot)$ is Pareto-optimal, then there exist a $j \in \{1,2,\ldots,N\}$ and N-1 real numbers r_i such that

$$V_j(x_0,u^0(\cdot)) \leqq V_j(x_0,u(\cdot))$$

for all $u(\cdot)$ playable at x_0 and subject to

$$V_i(x_0,u(\cdot)) \leqq r_i \quad , \quad i \neq j \quad , \quad i = 1,2,\ldots,N \ .$$

This lemma has the

Corollary 3.1. If $u^0(\cdot)$ is Pareto-optimal, then for all $j \in \{1,2,\ldots,N\}$

$$V_j(x_0,u^0(\cdot)) \leqq V_j(x_0,u(\cdot))$$

for all $u(\cdot)$ playable at x_0 and subject to

$$V_i(x_0,u(\cdot)) \leqq V_i(x_0,u^0(\cdot)) \quad , \quad i \neq j \quad , \quad i = 1,2,\ldots,N \ .$$

3.3 Sufficient Conditions

By means of the following, easily proven lemmas (e.g., Ref. 6) one can reduce sufficient conditions for Pareto-optimal controls to sufficient conditions for optimal control (e.g., Ref. 10).

Lemma 3.2. Control N-tuple $u^0(\cdot)$, playable at x_0, is Pareto-optimal if there is an $\alpha \in R^N$, $\alpha_i > 0$, $i = 1,2,\ldots,N$, such that

$$\sum_{i=1}^{N} \alpha_i V_i(x_0,u^0(\cdot)) \leqq \sum_{i=1}^{N} \alpha_i V_i(x_0,u(\cdot))$$

for all $u(\cdot)$ playable at x_0.

Lemma 3.3.+ Control N-tuple $u^0(\ )$, playable at x_0, is Pareto-optimal if there is an $\alpha \in R^N$, $\alpha_i \geqq 0$, $i=1,2,\ldots,N$, and $\sum_{i=1}^{N} \alpha_i = 1$, such that

---

+We say that $u(\cdot) \neq u^0(\cdot)$ iff $u(\tau) \neq u^0(\tau)$ on a subset of $[0,\tau_1]$ having positive measure.

$$\sum_{i=1}^{N} \alpha_i V_i(x_0, u^0(\cdot)) < \sum_{i=1}^{N} \alpha_i V_i(x_0, u(\cdot))$$

for all $u(\cdot) \neq u^0(\cdot)$ playable at x_0.

4. NON-COOPERATIVE DIFFERENTIAL GAMES

4.1 Nash Equilibrium

For a dynamical system, Definition 1.5 becomes

Definition 4.1. A strategy N-tuple $p^e(\cdot)$ is an equilibrium on $X \subseteq R^n$ iff

(i) it is playable at all $x_0 \in X$, and for all $i \in \{1,2,\dots,N\}$ and $x_0 \in X$

(ii) $V_i(x_0, p^e(\cdot), x^e(\cdot)) \leq V_i(x_0, {}^ip(\cdot), x^i(\cdot))$ for all terminating plays $\{x_0, p^e(\cdot), x^e(\cdot)\}$ and $\{x_0, {}^ip(\cdot), x^i(\cdot)\}$, where

$${}^ip(\cdot) = \{p^{1e}(\cdot), \dots, p^{i-1e}(\cdot), p^i(\cdot), p^{i+1e}(\cdot), \dots, p^{Ne}(\cdot)\} .$$

For two-person zero-sum games

$$V_1(x_0, p(\cdot), x(\cdot)) = - V_2(x_0, p(\cdot), x(\cdot)) \triangleq V(x_0, p(\cdot), x(\cdot))$$

and Definition 1.6 is similarly altered. In view of playability requirement (i), decision N-tuples in differential games are not members of a product of decision sets; hence, some of the desirable properties of classical two-person zero-sum games, listed in Section 1.3, need not hold (e.g., Refs. 11-12).

4.2 Necessary and Sufficient Conditions.

Necessary conditions and sufficient conditions for equilibrium strategies are rather lengthy and involved. For two-person zero-sum games, extensive discussions can be found in Refs. 6, 13 - 17, among others. For N-person nonzero-sum games, see Refs. 6, 18 - 19.

Suffice it to say that, for sufficiently well-behaved equilibria, there are necessary conditions akin to the Minimum Principle of optimal control theory. However, because of the use of closed-loop strategies in place of open-loop controls, the adjoint (costate) equations contain terms which involve the equilibrium strategies. In two-person zero-sum games, it is often possible to suppress these troublesome terms, for instance, when the constraints are state-independent and satisfy certain constraint qualifications; this is not possible, in general, for non-zero-sum games. Thus, it is well-nigh impossible to utilize necessary conditions in a constructive fashion to deduce candidates for equilibrium strategies. One exception is the class of so-called "trilinear games"[20] for which one can deduce a Nash equilibrium strategy N-tuple that depends only on time; such games occur in certain problems of micro-economics[21]. Another interesting class of equilibria arises in certain problems of bargaining (e.g., Section 5.2 and Refs. 22 - 23); these are equilibria for which condition (ii) of Definition 4.1 is met trivially.

5. EXAMPLES

5.1 Cooperative Bargaining during Strike

Consider a bargaining process during a strike whose state at time τ is defined by the offer by management, $x(\tau)$, and the demand by labor, $y(\tau)$, with initial demand exceeding offer, $x(0) = x_0 < y(0) = y_0$. The strike ends at time T, the first time $y(T) - x(T) = m > 0$; that is, labor accepts an offer when it is sufficiently close to its demand.

Both parties desire a quick end to the strike, but labor wants to minimize the demand at settlement time, whereas management wishes to minimize the final offer. Thus, the costs to management and labor, respectively are

$$V_1 = k_1 T + x(T) \text{ and } V_2 = k_2 T - y(T)$$

k_1, k_2 = constant > 0.

We suppose that the rates of concessions are proportional to the difference between demand and offer, and that the parties control the rates, Thus,

$$\dot{x}(\tau) = u(\tau)(y(\tau) - x(\tau)) \quad , \quad x(0) = x_0 \quad ,$$
$$\dot{y}(\tau) = -v(\tau)\ (y(\tau) - x(\tau)) \ , \quad y(0) = y_0$$

and

$$0 \leq u(\tau) \leq a, \quad 0 \leq v(\tau) \leq b$$

where a and b are the maximum values of management's and labor's controls, $u(t)$ and $v(t)$, respectively.

This bargaining process is considered as a cooperative game in Ref. 20 and all Pareto-optimal controls are deduced.

Consider two parameters, α_1 and α_2 with $\alpha_1 + \alpha_2 = 1$, and let $\beta = \alpha_1 - \alpha_2$ and $k = \alpha_1 k_1 + \alpha_2 k_2$. The set of Pareto-optimal controls $\{u^0(\cdot), v^0(\cdot)\}$ is found to be the following. For $\alpha_1 > \alpha_2$ (management "more important" than labor), we have two cases:

Case I. $k/\beta b \geq m$

$$u^0(\tau) = \begin{cases} 0 \text{ for } y(\tau) - x(\tau) \geq k/\beta b \ , \\ a \text{ for } y(\tau) - x(\tau) < k/\beta b \ , \end{cases} \qquad v^0(\tau) \equiv b$$

Case II. $k/\beta b < m$

$$u^0(\tau) \equiv 0, \quad v^0(\tau) \equiv b$$

The time at which management begins to make concessions (Case I) is

$$T^0 = \frac{1}{b} \ln \frac{\beta b(y_0 - x_0)}{k}$$

This time does not depend on management's control limit a; however, the larger labor's control limit b (that is, the more rapidly labor is willing to concede), the later management begins to concede. If either b or m is sufficiently large, so that $k/\beta b < m$ (that is, labor makes concessions "rapidly" or is willing to settle when offer falls "appreciably" below demand), then management does not concede at all.

For $\alpha_1 < \alpha_2$ (labor "more important" than management), the situation outlined above obtains with the roles of management and labor exchanged. Finally, if $\alpha_1 = \alpha_2$ (both parties "equally important"), both concede at maximum rate; that is,

$$u^0(\tau) \equiv a, \quad v^0(\tau) \equiv b.$$

5.2 Non-cooperative Bargaining during Strike

As in Section 5.1, we consider a bargaining process during a strike. However, here we allow bargaining on more than one issue. That is $x(\tau)$ and $y(\tau)$ are n-vectors, being offer and demand, respectively, with $x(0) = x_0$, $y(0) = y_0$, and $x_{0i} < y_{0i}$, $i = 1,2,\ldots,n$. Settlement occurs at a time T, the first time $x(T) = y(T)$.

Again, we take the costs for management and labor to be[+]

$$V_1 = k_1 T + \ell_1' x(T)$$
$$V_2 = k_2 T - \ell_1' y(T)$$

[+] Prime denotes transpose.

where k_1 and k_2 are positive constants, and ℓ_1 and ℓ_2 are n-vectors with positive components.

We denote the concession rates by*

$$\dot{x}(\tau) = u(\tau)$$

$$\dot{y}(\tau) = -v(\tau)$$

and consider a non-cooperative situation. That is, among a class of admissible strategy pairs $\{\mu(\cdot), \nu(\cdot)\}$, where++

$$\mu(x(\tau),y(\tau)) = u(\tau)$$

$$\nu(x(\tau),y(\tau)) = v(\tau)$$

we seek a Nash equilibrium pair $\{\mu^e(\cdot),\nu^e(\cdot)\}$ on $\{(x,y) \in R^{2n} : 0 < y_i - x_i < \infty\}$.

It is readily verified that there is an equilibrium strategy pair

$$\mu^e(x,y) = \frac{k_2 z}{\ell_2' z} ,$$

$$\nu^e(x,y) = \frac{k_1 z}{\ell_1' z} ,$$

where $z = y - x$. For this equilibrium, condition (ii) of Definition 4.1 is met trivially. Thus, by using his equilibrium strategy, a player fixes his opponent's cost.

---

*This problem is treated in Ref. 20 where, however, it is assumed *a priori* that the concession rates depend on $y(\tau) - x(\tau)$. A generalization of this problem is in Ref. 21.

++As can readily be shown, the strategies do not depend on time.

In addition, we can draw the following conclusions:

(i) All issues are settled simultaneously.

(ii) The strike duration is $T^e = \dfrac{||z_0||}{c}$, where $z_0 = y_0 - x_0$,

$$c = \frac{k_1}{\ell_1' e} + \frac{k_2}{\ell_2' e} \text{ and } e = \frac{z_0}{||z_0||} .$$

(iii) The concession rate is given by $\dfrac{d||z||}{d\tau} = -c$.

REFERENCES

1. Vincent, T.L. and Leitmann, G., Control space properties of cooperative games, *J. of Optimization Theory and Appl.*, 6, 91, 1970.

2. Leitmann, G., Rocklin, S. and Vincent, T.L., A note on control space properties of cooperative games, *J. of Optimization Theory and Appl.*, 9, 379, 1972.

3. Yu, P.L. and Leitmann, G., Compromise solutions, domination structures and Salukvadze's solution, *J. of Optimization Theory and Appl.*, 13, 362, 1974.

4. Pareto, V., *Manuel d'economique politique*, Girard et Briere, Paris, 1909.

5. Nash, J., Non-cooperative games, *Annals of Mathematics*, 54, 286, 1951.

6. Leitmann, G., *Cooperative and non-cooperative many player differential games*, Springer Verlag, Vienna, 1974.

7. Schmitendorf, Wm. and Leitmann, G., *A simple derivation of necessary conditions for pareto optimality*, IEEE Transactions on Automatic Control, AC-19, 601, 1974.

8. Hestenes, M., *Calculus of variations and optimal control theory*, John Wiley and Sons, N.Y., 1966.

9. Schmitendorf, Wm., Pontryagin's principle for problems with isoperimetric constraints and for problems with inequality terminal constraints, *J. of Optimization Theory and Appl.*, to appear.

10. Leitmann, G. and Schmitendorf, Wm., Some sufficiency conditions for pareto-optimal control, *J. of Dynamic Systems, Measurement, and*

Control, 95, 356, 1973.

11. Leitmann, G and Rocklin, S., The effect of playability in differential games on the relation between best guaranteed costs and saddlepoint values, *J. of Optimization Theory and Appl.*, to appear.

12. Leitmann, G., On some consequences of playability in differential games, *Proceed. IEEE Conf. on Decision and Control*, 346, 1974.

13. Isaacs, R., *Differential Games*, John Wiley and Sons, N.Y., 1965.

14. Blaquière, A. and Leitmann, G., *Jeux Quantitatifs*, Gauthier-Villars, Paris, 1969.

15. Blaquière, A., Gérard, F., and Leitmann, G., *Quantitative and Qualitative Games*, Academic Press, N.Y., 1969.

16. Friedman, A., *Differential Games*, John Wiley and Sons, N.Y., 1971.

17. Berkovitz, L.D., Necessary conditions for optimal strategies in a class of differential games and control problems, *SIAM J. on Control*, 5, 1, 1967.

18. Case, J.H., Toward a theory of many player differential games, *SIAM J. on Control*, 7, 179, 1969.

19. Stalford, H. and Leitmann, G., Sufficiency conditions for Nash equilibria in N-person differential games, in *Topics in differential games* (ed. A. Blaquière), North-Holland Publ. Co., Amsterdam, 345, 1973.

20. Clemhout, S. and Wan, H. Y. Jr., A class of trilinear games, *J. of Optimization Theory and Appl.*, 14, 419, 1974.

21. Leitmann, G. and Liu, P. T., A differential game model of labor-management negotiation during a strike, *J. of Optimization Theory*

and Appl., 13, 427, 1974 and 14, 443, 1974.

22. Clemhout, S., Leitmann, G. and Wan, H. Y. Jr., A model of bargaining under strike: A differential game view, *J. of Economic Theory*, to appear.

ADDITIONAL BIBLIOGRAPHY

Zadeh, L. A., Optimality and non-scalar-valued performance criteria, *IEEE Trans. Autom. Cont.*, AC-8, 59, 1963.

Da Cunha, N. O. and Polak, E., Constrained minimization under vector-valued criteria in linear topological spaces, in *Mathematical Theory of Control* (eds., Balakrishnan, A. V. and Neustadt, L. W.), Academic Press, N.Y., 1967.

Ho, Y. C. and Leitmann, G., Eds., *Proceedings of the First International Conference on the Theory and Application of Differential Games*, Amherst, Mass., 1969.

Starr, A. W. and Ho, Y. C., Nonzero-sum differential games, *J. of Optimization Theory and Appl.*, 3, 184, 1969.

Starr, A. W., and Ho, Y.C., Further properties of nonzero-sum differential games, *J. of Optimization Theory and Appl.*, 3, 207, 1969.

Ho., Y. C., Differential games, dynamic optimization and generalized control theory, *J. of Optimization Theory and Appl.*, 6, 179, 1970.

Ciletti, M. D., New results in the theory of differential games with information time lag, *J. of Optimization Theory and Appl.*, 8, 287, 1971.

Clemhout, S., Leitmann, G. and Wan, H. Y. Jr., A differential game model of duopoly, *Econometrica*, 39, 911, 1971.

Blaquière, A., Juricek, L. and Wiese, K., Geometry of pareto equilibria and a maximum principle in N-person differential games, *J. of Math. Analysis and Appl.*, 38, 223, 1972.

Athans, M. and Geering, H. P., Necessary and sufficient conditions for a differentiable nonscalar-valued functions to attain extrema, *IEEE Trans. Autom. Contr.*, AC-18, 132, 1973.

Clemhout, S., Leitmann, G. and Wan, H. Y., Jr., A differential game model of oligopoly, *J. of Cybernetics*, 3, 24, 1973.

Haurie, A., On pareto optimal decisions for a coalition of a subset of players, *IEEE Trans. Autom. Contr.*, AC-18, 144, 1973.

Leitmann, G., Collective bargaining: A differential game, *J. of Optimization Theory and Appl.*, 11, 405, 1973.

Schmitendorf, W. E., Cooperative games and vector-valued criteria problems, *IEEE Trans. Autom. Contr.*, AC-18, 139, 1973.

Simaan, M. and Cruz, J. B., On the Stackelberg strategy in nonzero-sum games, *J. of Optimization Theory and Appl.*, 11, 533, 1973.

Ciletti, M. D., Canonical equations for the generalized Hamilton-Jacobi equation in DGWITL, *J. of Optimization Theory and Appl.*, 13, 262, 1974.

Haurie, A. and Delfour, M. C., Individual and collective rationality in a dynamic pareto equilibrium, *J. of Optimization Theory and Appl.*, 13, 290, 1974.

Ho, Y. C., On the minimax principle and zero-sum stochastic differential games, *J. of Optimization Theory and Appl.*, 13, 343, 1974.

Mori, K. and Shimemura, E., Linear differential games with delayed and noisy information, *J. of Optimization Theory and Appl.*, 13, 275, 1974.

Wilson, D. J., Mixed strategy solutions for quadratic games, *J. of Optimization Theory and Appl.*, 13, 319, 1974.

Yu, P. L. and Leitmann, G., Nondominated decisions and cone convexity in dynamic multicriteria decision problems, *J. of Optimization Theory and Appl.*, 14, 573, 1974.

VECTOR-VALUED OPTIMIZATION IN MULTI-PLAYER QUANTITATIVE GAMES

by

A. BLAQUIERE,

Laboratoire d'Automatique Théorique
Université de Paris 7
Paris, France

A. GLOBAL PROPERTIES OF A GAME SURFACE

1. Vector Valued Optimization

First, let us define a N-player quantitative game as in Ref. (1).

Let there be given a set G, whose members will be denoted by x , ordered by a reflexive relation $\succeq$, such that x' $\succeq$ x" or x" $\succeq$ x' whenever x' and x" are distinct members of the union of the domain and the range of $\succeq$. Let there be given prescribed sets S_α, $\alpha = 1,2,\ldots N$, whose members will be denoted by s_α , $\alpha = 1,2,\ldots N$, respectively ; and let there be given a relation $R \subset D \times P(G)$, where $D = G \times S_1 \times \cdots \times S_N$ and $P(G)$ is the collection of all non empty subsets of G . We shall suppose that $(x^i,s) R \gamma$, where $s = (s_1, s_2, \ldots s_N)$, $(x^i,s) \in D$, $\gamma \in P(G)$, implies that

(a) $x^i \in \gamma$, and

(b) $x \succeq x^i, \quad \forall x \in \gamma$

Members x of G are *states* of the game. Members s_α of S_α , $\alpha = 1,2,\ldots N$, are *strategies* of players J_α , $\alpha = 1,2,\dots N$, respectively. Members s of $S_1 \times \dots \times S_N$ are *strategy N-tuples*. Members γ of *Range* R are *trajectories*. If $(x^i,s)\,R\,\gamma$, we shall say that trajectory γ is *generated from* x^i *by strategy N-tuple* s . x^i is its *initial point*.

Now, in view of defining optimality of a strategy N-tuple, let there be given N reflexive and symmetric relations $C_\alpha \subset S^2$, $S = S_1 \times \dots \times S_N$, $\alpha = 1,2,\dots N$. If $s\,C_\alpha\,s'$ we shall say that strategy N-tuples s and s' are *comparable* for player J_α .

For instance, in a two-player game where optimality of a strategy pair is to be defined in the sense of Nash, we shall let

$$(s_1,s_2)\ C_1(s_1',s_2') \iff s_2 = s_2' \text{ , and}$$

$$(s_1,s_2)\ C_2(s_1',s_2') \iff s_1 = s_1'$$

that is, for each player the comparison is done for a fixed strategy, otherwise arbitrary, of the opponent player.

If $x^p \in \gamma$, $\gamma \in$ Range R, we shall let

$$\rho(\gamma,x^p) = \{\, x : \quad x \in \gamma , \quad x \succ x^p \,\}$$

and if x^p and x^q are members of γ such that $x^q \succ x^p$, we shall let

$$\rho(\gamma,x^p,x^q) = \{\, x : \quad x \in \gamma, \quad x^q \succ x \succ x^p \,\}$$

Let us introduce

Assumption 1. If $\gamma \in$ Range R and $x^p \in \gamma$, then

$$(x^i,s)\,R\,\gamma \Rightarrow \begin{cases} (x^i,s)\,R\,\rho(\gamma,x^i,x^p) \text{ and} \\ (x^p,s)\,R\,\rho(\gamma,x^p) \end{cases}$$

Assumption 2.

$$\left.\begin{array}{l} (x^i,s')\,R\,\gamma', \text{ and} \\ (x^j,s'')\,R\,\gamma'', \text{ and} \\ x^j \in \gamma' \end{array}\right\} \Rightarrow \begin{cases} \text{There exists } s \in S \text{ such that (a)} \\ (x^i,s)\,R\,(\rho(\gamma',x^i,x^j) \cup \gamma'') \end{cases}$$

and (b)

$$s'C_\alpha s'' \Rightarrow \begin{cases} sC_\alpha s' \text{ , and} \\ sC_\alpha s'' \end{cases} \qquad \alpha \in \{1, 2, \ldots N\}$$

and $s' = s'' \Rightarrow s = s' = s''$.

If there exists x^j, $x^j \in \gamma$, $\gamma \in$ Range R, such that $x^j \succ x$ for any $x \in \gamma$, then we shall say that x^j is the *end point* of γ . Let there be given a subset θ of G, namely the *target set*. A trajectory γ will be called a *path* and denoted by $\bar{\gamma}$ if none of its members, with the possible exception of its end point when it exists, belong to θ.

Finally, let there be given N linear spaces Ω_α, $\alpha = 1,2,\ldots N$, reflexive relations $(\geq)_\alpha$, $(\geq)_\alpha \subset \Omega_\alpha^2$, $\alpha = 1,2,\ldots N$, and mappings V_α, $\alpha = 1,2,\ldots N$:

$$V_\alpha \begin{cases} R \to \Omega_\alpha \\ (x^i,s,\gamma) \mapsto V_\alpha(x^i,s,\gamma) \end{cases} \qquad \alpha = 1,2,\ldots N$$

V_α is the *performance index* of player J_α, and $V_\alpha(x^i,s,\gamma)$ is the *pay-off* of player J_α for a trajectory γ generated from x^i by strategy N-tuple s. $(\geq)_\alpha$ is the *preference relation*† of player J_α .

We shall say that a strategy N-tuple s is *playable* at state x^i if there exists a path $\bar{\gamma}$ such that $(x^i,s) R \bar{\gamma}$ and $\bar{\gamma} \cap \theta \neq \phi$. We shall denote by $J(x^i)$ the set of all strategy N-tuples playable at state x^i. A path $\bar{\gamma}$ such that $\bar{\gamma} \cap \theta \neq \phi$ is called a *terminating path*. We shall denote by $I(x^i,s,\theta)$ the set of all terminating paths generated from a state x^i by a strategy N-tuple s .

---

† A preference relation need not be transitive.

Then, we shall say that a strategy N-tuple s^* is *C-optimal at state* x^i, $C = (C_1, C_2, \dots C_N)$, if

(i) $s^* \in J(x^i)$, and

(ii) there exists $\omega = (\omega_1, \omega_2, \dots \omega_N) \in \Omega$, $\Omega = \Omega_1 \times \dots \times \Omega_N$, such that

$$V(x^i, s^*, \gamma^*) = \omega \qquad \forall \gamma^* \in I(x^i, s^*, \theta)$$

where $V = (V_1, V_2, \dots V_N)$, and

$V^*(x^i) = (V_1^*(x^i), V_2^*(x^i), \dots V_N^*(x^i)) = \omega$ is the *value of the game at state* x^i ; and

(iii) $V_\alpha^*(x^i) (\geqslant)_\alpha V_\alpha(x^i, s, \bar{\gamma})$

$\forall s$ such that $s \in J(x^i)$ and $s C_\alpha s^*$ and $\forall \bar{\gamma} \in I(x^i, s, \theta)$ $\Big\}$ $\alpha = 1, 2, \dots N$

We shall denote by $J^*(x^i)$ the set of all strategy N-tuples C-optimal at state x^i. At last a N-player quantitative game is defined by $(G, \mathcal{E}, S, R, C, \Omega, \{(\geqslant)_\alpha, \alpha = 1, 2, \dots N\}, V, \theta)$.

In addition to Assumptions 1 and 2 we shall introduce

Assumption 3. For any (x^i, s', γ') and (x^j, s'', γ'') such that $(x^i, s') R \gamma'$ and $(x^j, s'') R \gamma''$ and x^j is the end point of γ'

$$V(x^i, s', \gamma') + V(x^j, s'', \gamma'') = V(x^i, s, \gamma)$$

where $\gamma = \gamma' \cup \gamma''$ and s is a strategy N-tuple such that $(x^i, s) R \gamma$; and condition (b) of Assumption 2 is satisfied.

2. Surface of the game

In the following we shall suppose that there exists a strategy N-tuple s^* which is C-optimal at state x^i, and we shall denote by X the set of all states x at which s^* is C-optimal.

We shall define $(>)_\alpha$, $\alpha \in \{1, 2, \dots N\}$, by

$$\omega'_\alpha (>)_\alpha \omega''_\alpha \Leftrightarrow (\omega'_\alpha (\geqslant)_\alpha \omega''_\alpha \text{ and } \textit{not } \omega''_\alpha (\geqslant)_\alpha \omega'_\alpha), \quad (\omega'_\alpha, \omega''_\alpha) \in \Omega_\alpha^2$$

Then we shall let

$$\omega' > \omega'' \iff \omega'_\alpha (>)_\alpha \, \omega''_\alpha , \qquad \alpha = 1,2,\ldots N$$

$$\omega' = (\omega'_1, \omega'_2, \ldots \omega'_N) \in \Omega, \qquad \omega'' = (\omega''_1, \omega''_2, \ldots \omega''_N) \in \Omega .$$

Now let us define the set $\Sigma(C)$, $C \in \Omega$, by

$$\Sigma(C) = \{y = (x_o,x) : \ (x_o,x) \in \Omega \times G, \ \ x_o + V^*(x) = C\}$$

$\Sigma(C)$ is the *surface of the game* for given C.

Let us define also a set $\Sigma_\alpha(C_\alpha)$, $C_\alpha \in \Omega_\alpha$, namely

$$\Sigma_\alpha(C_\alpha) = \{y_\alpha = (x_{o\alpha},x) : \ (x_{o\alpha},x) \in \Omega_\alpha \times G, \ x_{o\alpha} + V^*_\alpha(x) = C_\alpha\}$$

$\alpha \in \{1,2,\ldots N\}$.

$\Sigma_\alpha(C_\alpha)$ will be called an α-*surface of the game*.

Note that, in the definitions of $\Sigma(C)$ and $\Sigma_\alpha(C_\alpha)$, V^* and V^*_α , $\alpha = 1,2,\ldots N$, are mappings $X \to \Omega$ and $X \to \Omega_\alpha$, $\alpha = 1,2,\ldots N$, respectively.

In connection with the definition of $\Sigma(C)$, $C = (C_1,C_2,\ldots C_N) \in \Omega$, we shall define the sets $A/\Sigma(C)$ and $\overline{B/\Sigma(C)}$ by

$$A/\Sigma(C) = \{y = (x_o,x) : \ (x_o,x) \in \Omega \times X, \ \ x_o + V^*(x) > C\}$$

$$\overline{B/\Sigma(C)} = \{y = (x_o,x) : \ (x_o,x) = (x_{o1}, x_{o2}, \ldots x_{oN}, x) \in \Omega \times X,$$
$$\exists \alpha \in \{1,2,\ldots N\}, \quad C_\alpha (\geqslant)_\alpha \, x_{o\alpha} + V^*_\alpha(x)\}$$

A point $y \in A/\Sigma(C)$ is called an A-point, and a point $y \in \overline{B/\Sigma(C)}$ is called a B-point, relative to $\Sigma(C)$.

In connection with the definition of $\Sigma_\alpha(C_\alpha)$, $C_\alpha \in \Omega_\alpha$, we shall define the sets $A/\Sigma_\alpha(C_\alpha)$ and $\overline{B/\Sigma_\alpha(C_\alpha)}$, namely

$$A/\Sigma_\alpha(C_\alpha) = \{y_\alpha = (x_{o\alpha},x) : \ (x_{o\alpha},x) \in \Omega_\alpha \times X, \ \ x_{o\alpha} + V^*_\alpha(x)(>)_\alpha \, C_\alpha\}$$

$$\overline{B/\Sigma_\alpha(C_\alpha)} = \{y_\alpha = (x_{o\alpha},x) : \ (x_{o\alpha},x) \in \Omega_\alpha \times X, \ \ C_\alpha (\geqslant)_\alpha \, x_{o\alpha} + V^*_\alpha(x)\}$$

$\alpha \in \{1,2,\ldots N\}$.

A point $y_\alpha \in A/\Sigma_\alpha(C_\alpha)$ is called an A-point, and a point $y_\alpha \in \overline{B/\Sigma_\alpha(C_\alpha)}$ is called a B-point, relative to $\Sigma_\alpha(C_\alpha)$.

At last, we shall consider the mappings

$$\Phi_\alpha \begin{cases} \Omega_\alpha \times X \to \Omega_\alpha \\ y_\alpha = (x_{o\alpha}, x) \mapsto \Phi_\alpha(y_\alpha) = x_{o\alpha} + V^*_\alpha(x) \end{cases}$$

and

$$\Phi \begin{cases} \Omega \times X \to \Omega \\ y = (x_o, x) \mapsto \Phi(y) = x_o + V^*(x) \end{cases}$$

and we shall denote the transforms of $A/\Sigma_\alpha(C_\alpha)$ and $\overline{B/\Sigma_\alpha(C_\alpha)}$, under the mapping Φ_α, by $A(C_\alpha)$ and $\overline{B(C_\alpha)}$, respectively. Likewise, we shall denote the transforms of $A/\Sigma(C)$ and $\overline{B/\Sigma(C)}$, under the mapping Φ, by $A(C)$ and $\overline{B(C)}$, respectively.

From now on, we shall let $Z = \Omega \times G$, and $Z^* = \Omega \times X$, and $Z_\alpha = \Omega_\alpha \times G$, and $Z^*_\alpha = \Omega_\alpha \times X$, $\alpha = 1,2,\ldots N$.

3. Trajectories and paths in $P(Z)$†

For any $(x^i, s) \in D$ and for any $\gamma \in$ Range R such that $(x^i, s) R \gamma$, and for any $C \in \Omega$, we shall define a *trajectory* $\Gamma(\gamma, y^i, s)$ *in* $P(Z)$ by

$$\Gamma(\gamma, y^i, s) = \{y = (x_o, x):\ \ x_o \in \Omega,\ \ x \in \gamma,\ \ x_o + V(x, s, \rho(\gamma, x)) = C\}$$

$$y^i = (x^i_o, x^i), \qquad C = x^i_o + V(x^i, s, \gamma)$$

We shall say that $\Gamma(\gamma, y^i, s)$ is generated from initial point y^i by strategy N-tuple s . A trajectory $\Gamma(\gamma, y^i, s)$ will be called a *path in* $P(Z)$ if γ is a path. It will be called a *terminating path in* $P(Z)$ if γ is a terminating path, in which case $\Gamma(\gamma, y^i, s)$ reaches the set $\Theta = \Omega \times \theta$.

We shall define an α-*trajectory* $\Gamma_\alpha(\gamma, y^i_\alpha, s)$ *in* $P(Z_\alpha)$†† by

† $P(Z)$ is the collection of all non empty subsets of Z

†† $P(Z_\alpha)$ is the collection of all non empty subsets of Z_α

$\Gamma_\alpha(\gamma, y^i_\alpha, s) = \{y_\alpha = (x_{o\alpha}, x): \ x_{o\alpha} \in \Omega_\alpha,\ x \in \gamma,\ x_{o\alpha} + V_\alpha(x, s, \rho(\gamma, x)) = C_\alpha\}$

$y^i_\alpha = (x^i_{o\alpha}, x^i) \qquad C_\alpha = x^i_{o\alpha} + V_\alpha(x^i, s, \gamma)$

Again, we shall say that $\Gamma_\alpha(\gamma, y^i_\alpha, s)$ is generated from initial point y^i_α by strategy N-tuple s, and $\Gamma_\alpha(\gamma, y^i_\alpha, s)$ will be called an *α-path in* $P(Z_\alpha)$ if γ is a path.

We shall say that a path $\Gamma(\gamma, y^i, s)$, or an α-path $\Gamma_\alpha(\gamma, y^i_\alpha, s)$, is *optimal* if $s = s^* \in J^*(x^i)$ and $\gamma = \gamma^* \in I(x^i, s^*, \theta)$.

4. Lemma 1.

Let us introduce

Assumption 4. For any ω_α, $\omega_\alpha \in \Omega_\alpha$, $\alpha \in \{1,2,\dots N\}$,

$$(\omega'_\alpha + \omega_\alpha)(\geq)_\alpha(\omega''_\alpha + \omega_\alpha) \iff \omega'_\alpha(\geq)_\alpha\omega''_\alpha$$

Then we have

Lemma 1. If s^* is C-optimal at state x^i, and $\gamma^* \in I(x^i, s^*, \theta)$, and $x^j \in \gamma^*$, then s^* is C-optimal at state x^j.

Proof. Clearly, s^* is playable at state x^j. Let $\overline{\gamma''} \in I(x^j, s^*, \theta)$ and $\overline{\gamma'} = \rho(\gamma^*, x^i, x^j)$ and $\bar\gamma = \overline{\gamma'} \cup \overline{\gamma''}$. From Assumption 2 we have $(x^i, s^*)R\bar\gamma$, and indeed $\bar\gamma \in I(x^i, s^*, \theta)$. From Assumption 3 and condition (ii) regarding s^*, we have

$V(x^i, s^*, \bar\gamma) = V^*(x^i) = V(x^i, s^*, \overline{\gamma'}) + V(x^j, s^*, \overline{\gamma''})$, and

$V(x^i, s^*, \gamma^*) = V^*(x^i) = V(x^i, s^*, \overline{\gamma'}) + V(x^j, s^*, \rho(\gamma^*, x^j))$

and accordingly

$$V(x^j, s^*, \overline{\gamma''}) = V(x^j, s^*, \rho(\gamma^*, x^j))$$

Now let $\overline{\gamma''} \in I(x^j, s'', \theta)$ and $s'' C_\alpha s^*$ for $\alpha \in \{1,2,\dots N\}$. Let $\overline{\gamma'} = \rho(\gamma^*, x^i, x^j)$ and $\bar\gamma = \overline{\gamma'} \cup \overline{\gamma''}$. From Assumption 2, there exists a strategy N-tuple s, $sC_\alpha s^*$, such that $(x^i, s)R\bar\gamma$, and indeed $\bar\gamma \in I(x^i, s, \theta)$. From Assumption 3 we have

$V_\alpha(x^i,s,\bar{\gamma}) = V_\alpha(x^i,s^*,\overline{\gamma'}) + V_\alpha(x^j,s'',\overline{\gamma''})$; and

$V_\alpha(x^i,s^*,\gamma^*)=V_\alpha(x^i,s^*,\overline{\gamma'}) + V_\alpha(x^j,s^*,\rho(\gamma^*,x^j))$

At last, from Condition (iii) and Assumption 4, we obtain

$V_\alpha(x^j,s^*,\rho(\gamma^*,x^j))(\geqslant)_\alpha V_\alpha(x^j,s'',\overline{\gamma''}), \quad \alpha \in \{1,2,\ldots N\}$.

Hence Lemma 1 is proved.

5. A fundamental property of game surfaces

A fundamental property of game surfaces is embodied in

Theorem 1. No point of an α-path $\Gamma_\alpha(\bar{\gamma},y^i_\alpha,s)$, generated from $y^i_\alpha = (x_{o\alpha},x^i)$ by strategy N-tuple s, such that $(x^i,s)\,R\bar{\gamma}$, $sC_\alpha s^*$, $s^* \in J^*(x^i)$, $\alpha \in \{1,2,\ldots N\}$, is an A-point relative to the α-surface of the game through y^i_α ; and, furthermore, if $y^j_\alpha = (x^j_{o\alpha},x^j) \in \Gamma_\alpha(\bar{\gamma},y^i_\alpha,s) \cap Z^*_\alpha$ then y^j_α is a B-point relative to that α-surface.

Proof. Let $x^j \in \bar{\gamma}$. If $s^* \notin J^*(x^j)$, then $x^j \notin X$ and the conclusion of Theorem 1 is trivial. Accordingly let us suppose that $s^* \in J^*(x^j)$ and let $\gamma^* \in I(x^j,s^*,\theta)$. We have

$$V_\alpha(x^j,s^*,\gamma^*) = V^*_\alpha(x^j)$$

Let $\overline{\gamma'} = \rho(\bar{\gamma},x^i,x^j)$ and $\overline{\gamma''} = \overline{\gamma'} \cup \gamma^*$. From Assumption 2, there exists a strategy N-tuple s'' such that $(x^i,s'')\,R\overline{\gamma''}$ and $s''C_\alpha s^*$. From Assumption 3 we have

$$V_\alpha(x^i,s'',\overline{\gamma''}) = V_\alpha(x^i,s,\overline{\gamma'}) + V^*_\alpha(x^j)$$

and from Condition (iii) regarding s^*, since $s''C_\alpha s^*$, we have

$$V^*_\alpha(x^i)(\geqslant)_\alpha V_\alpha(x^i,s'',\bar{\gamma})$$

Then, from Assumption 4 we obtain

$$V^*_\alpha(x^i) - V^*_\alpha(x^j)(\geqslant)_\alpha V_\alpha(x^i,s,\overline{\gamma'}) \qquad (1)$$

Now, let $y^j_\alpha = (x^j_{o\alpha},x^j) \in \Gamma_\alpha(\bar{\gamma},y^i_\alpha,s)$, and

$$C'_\alpha = x^i_{o\alpha} + V_\alpha(x^i,s,\overline{\gamma})$$

From the definition of $\Gamma_\alpha(\overline{\gamma},y^i_\alpha,s)$ we have

$$C'_\alpha = x^j_{o\alpha} + V_\alpha(x^j,s,\rho(\overline{\gamma},x^j)) \tag{2}$$

and from Assumption 3 it follows that

$$x^i_{o\alpha} + V_\alpha(x^i,s,\overline{\gamma'}) + V_\alpha(x^j,s,\rho(\overline{\gamma},x^j)) = C'_\alpha \tag{3}$$

From (2) and (3) we have

$$V_\alpha(x^i,s,\overline{\gamma'}) = x^j_{o\alpha} - x^i_{o\alpha}$$

and (1) rewrites

$$V^*_\alpha(x^i) - V^*_\alpha(x^j)(\geqslant)_\alpha x^j_{o\alpha} - x^i_{o\alpha} \tag{4}$$

From (4) and Assumption 4 we deduce

$$x^i_{o\alpha} + V^*_\alpha(x^i)(\geqslant)_\alpha x^j_{o\alpha} + V^*_\alpha(x^j) \tag{5}$$

Let $\Sigma_\alpha(C_\alpha)$ be the α-surface of the game through y^i_α, we have

$$x^i_{o\alpha} + V^*_\alpha(x^i) = C_\alpha$$

so that (5) rewrites

$$C_\alpha(\geqslant)_\alpha\, x^j_{o\alpha} + V^*_\alpha(x^j) \tag{6}$$

It follows that

$$y^j_\alpha \in \overline{B/\Sigma_\alpha(C_\alpha)}$$

which in turn implies that

$$y^j_\alpha \notin A/\Sigma_\alpha(C_\alpha)$$

Hence Theorem 1 is proved.

Corollary 1. No point of a path $\Gamma(\overline{\gamma},y^i,s)$ generated from $y^i = (x^i_o,x^i)$ by strategy N-tuple s, such that $(x^i,s)\,\mathcal{R}\,\overline{\gamma}$, $sC_\alpha s^*$, $s^* \in J^*(x^i)$, $\alpha \in \{1,2,\dots N\}$ is an A-point relative to the game surface through y^i, and, furthermore, if $y^j \in \Gamma(\overline{\gamma},y^i,s) \cap Z^*$ then y^j is a B-point relative to that game surface.

Corollary 1 is a direct consequence of Theorem 1 and the definition of $A/\Sigma(C)$ and $\overline{B/\Sigma(C)}$.

Theorem 2. An optimal α-path in $P(Z_\alpha)$ emanating from y^i_α has all of its points in the α-surface of the game through y^i_α .

Theorem 2 is a direct consequence of Lemma 1.

Corollary 2. An optimal path in $P(Z)$ emanating from y^i has all of its points in the game surface through y^i.

Corollary 2 is a direct consequence of Theorem 2 and the definition of a game surface.

B. LOCAL PROPERTIES OF A GAME SURFACE

6. Contingent of a set

Let Q be a non empty subset of a normed linear space E. We shall say that t, $t \in E$, is *tangent* to Q at z, $z \in E$. if the following condition is fulfilled.

There exists an infinite sequence in E,

$\tau = \{t^\nu : \nu = 1,2,\dots \text{ and } \|t^\nu - t\| \to 0 \quad \text{as} \quad \nu \to \infty\}$

and an infinite sequence of strictly positive numbers

$\varepsilon = \{\varepsilon^\nu : \nu = 1,2,\dots \text{ and } \varepsilon^\nu \to 0 \quad \text{as} \quad \nu \to \infty\}$

such that

$$z + \varepsilon^\nu t \in Q \qquad \nu = 1,2,\dots$$

The *contingent* of Q at z is the set $T(Q,z)$

$T(Q,z) = \{ z + t : \ t \text{ is tangent to } Q \text{ at } z \}$

7. Directional Preference on a contingent

Let $\geqslant$ be a preference relation on Q, and let $z \in Q$. If there exists an infinite sequence τ and an infinite sequence ε such that $z + \varepsilon^\nu t^\nu \in Q$ and $z + \varepsilon^\nu t^\nu \geqslant z$, $\nu = 1,2,\dots$, we shall let

$$t \dot{\geqslant} 0$$

and, if there exists τ and ε such that

$$z + \varepsilon^\nu t^\nu \in Q \quad \text{and} \quad z \geqslant z + \varepsilon^\nu t^\nu, \qquad \nu = 1,2,\ldots$$

we shall let

$$t \dot{\leqslant} 0$$

We shall say that $\dot{\geqslant}$ is a *directional preference induced by* $\geqslant$ on $T(Q,z)$.

8. Contingent of an α-surface of the game

Let us introduce

Assumption 5. Ω_α for $\alpha = 1,2,\ldots N$, and G, are normed linear spaces.

We shall say that a point $y_\alpha = (x_{o\alpha}, x) \in \Sigma_\alpha(C_\alpha)$, $\alpha \in \{1,2,\ldots N\}$, is an *interior point* of $\Sigma_\alpha(C_\alpha)$ if x is an interior point of X.

We shall say that y_α is *nice* if

(a) y_α is an interior point of $\Sigma_\alpha(C_\alpha)$; and

(b) there exists a positively homogeneous of order one, and continuous mapping

$$L_\alpha(x;\cdot) : \quad G \to \Omega_\alpha$$

and an open ball $B(x,r) \subset X$ of radius r and center x such that, for all $x + \varepsilon\eta \in B(x,r)$ where ε is a strictly positive number and $\eta \in G$,

$$V^*_\alpha(x + \varepsilon\eta) = V^*_\alpha(x) + L_\alpha(x\,;\,\varepsilon\eta) + o(\varepsilon,\eta)$$

where $\dfrac{\|o(\varepsilon,\eta)\|}{\varepsilon} \to 0$ uniformely in η as $\varepsilon \to 0$.

For $y_\alpha = (x_{o\alpha}, x) \in Z^*_\alpha$, $\alpha \in \{1,2,\ldots N\}$, we shall define $\operatorname{grad} \Phi_\alpha(y_\alpha)$ by

$$\operatorname{grad} \Phi_\alpha(y_\alpha) = (1,\ L_\alpha(x;\cdot))$$

and for any $t_\alpha = (\eta_{o\alpha}, \eta) \in Z_\alpha$ we shall let

$$\operatorname{grad} \Phi_\alpha(y_\alpha)\cdot t_\alpha = \eta_{o\alpha} + L_\alpha(x;\eta)$$

Lemma 2. If $y_\alpha = (x_{o\alpha}, x)$ is a nice point of $\Sigma_\alpha(C_\alpha)$, $C_\alpha \in \Omega_\alpha$,

$\alpha \in \{1,2,\ldots N\}$, then

$$T(\Sigma_\alpha(C_\alpha), y_\alpha) = \{y_\alpha + t_\alpha \ : \ \text{grad}\ \Phi_\alpha(y_\alpha)\cdot t_\alpha = 0\}$$

Proof. Let $t_\alpha = (\eta_{o\alpha}, \eta)$ be tangent to $\Sigma_\alpha(C_\alpha)$ at point y_α. Then, there exists an infinite sequence in Z_α

$$\{t_\alpha^\nu = (\eta_{o\alpha}^\nu, \eta^\nu) : \alpha = 1,2,\ldots \text{ and } \|t_\alpha^\nu - t_\alpha\| \to 0 \text{ as } \nu \to \infty\}$$

and an infinite sequence of strictly positive numbers

$$\{\varepsilon^\nu : \nu = 1,2,\ldots \text{ and } \varepsilon^\nu \to 0 \text{ as } \nu \to \infty\}$$

such that

$$x_{o\alpha} + \varepsilon^\nu\eta_{o\alpha}^\nu + V_\alpha^*(x + \varepsilon^\nu\eta^\nu) = C_\alpha \qquad \nu = 1,2,\ldots \tag{7}$$

It follows that

$$x_{o\alpha} + \varepsilon^\nu\eta_{o\alpha}^\nu + V_\alpha^*(x) + L_\alpha(x ; \varepsilon^\nu\eta^\nu) + o(\varepsilon^\nu, \eta^\nu) = C_\alpha, \quad \nu = 1,2,\ldots$$

where $\dfrac{\|o(\varepsilon^\nu,\eta^\nu)\|}{\varepsilon^\nu} \to 0$ as $\nu \to \infty$.

Since $x_{o\alpha} + V_\alpha^*(x) = C_\alpha$ we obtain

$$\varepsilon^\nu\eta_{o\alpha}^\nu + L_\alpha(x ; \varepsilon^\nu\eta^\nu) + o(\varepsilon^\nu,\eta^\nu) = 0$$

Dividing by ε^ν, then letting $\nu \to \infty$, we obtain

$$\eta_{o\alpha} + L_\alpha(x ; \eta) = \text{grad}\ \Phi_\alpha(y_\alpha)\cdot t_\alpha = 0 \tag{8}$$

Conversely, let (8) be satisfied and let $\{\varepsilon^\nu : \nu = 1,2,\ldots\}$ be some infinite sequence of strictly positive numbers such that $\varepsilon^\nu \to 0$ as $\nu \to \infty$. Since $L_\alpha(x;\cdot)$ is positively homogeneous of order one, we have

$$\varepsilon^\nu\eta_{o\alpha} + L_\alpha(x ; \varepsilon^\nu\eta) = 0, \qquad \nu = 1,2,\ldots$$

and since $x_{o\alpha} + V_\alpha^*(x) = C_\alpha$, we have

$$x_{o\alpha} + \varepsilon^\nu\eta_{o\alpha} + V_\alpha^*(x) + L_\alpha(x ; \varepsilon^\nu\eta) = C_\alpha, \quad \nu = 1,2,\ldots$$

It follows that, for ν sufficiently large,

$$x_{o\alpha} + \varepsilon^\nu\eta_{o\alpha} + V_\alpha^*(x + \varepsilon^\nu\eta) + o(\varepsilon^\nu,\eta) = C_\alpha, \quad \nu = 1,2,\ldots$$

where $\dfrac{\|o(\varepsilon^\nu,\eta)\|}{\varepsilon^\nu} \to 0$ as $\nu \to \infty$.

This condition can be rewritten

$x_{o\alpha} + \varepsilon^{\nu}(\eta_{o\alpha} + \frac{o(\varepsilon^{\nu},\eta)}{\varepsilon^{\nu}}) + V_{\alpha}^{*}(x + \varepsilon^{\nu}\eta) = C_{\alpha}$, $\nu = 1,2,\ldots$ and since

$(\eta_{o\alpha} + \frac{o(\varepsilon^{\nu},\eta)}{\varepsilon^{\nu}}, \eta) \to t_{\alpha}$ as $\nu \to \infty$, it follows that

$$y_{\alpha} + t_{\alpha} \in T(\Sigma_{\alpha}(C_{\alpha}), y_{\alpha})$$

which concludes the proof of Lemma 2.

Now, if $t_{\alpha} = (\eta_{o\alpha},\eta)$ is tangent to $\overline{B/\Sigma_{\alpha}(C_{\alpha})}$ at point $y_{\alpha} = (x_{o\alpha},x) \in \Sigma_{\alpha}(C_{\alpha})$, relation (7) is replaced by

$$C_{\alpha}(\geqslant)_{\alpha}x_{o\alpha} + \varepsilon^{\nu}\eta_{o\alpha}^{\nu} + V_{\alpha}^{*}(x + \varepsilon^{\nu}\eta^{\nu}), \qquad \nu = 1,2,\ldots \tag{9}$$

and from the definition of Φ_{α} we have

$$x_{o\alpha} + \varepsilon^{\nu}\eta_{o\alpha}^{\nu} + V_{\alpha}^{*}(x + \varepsilon^{\nu}\eta^{\nu}) = \Phi_{\alpha}(y_{\alpha} + \varepsilon^{\nu}t_{\alpha}^{\nu}) \in \overline{B(C_{\alpha})} \tag{10}$$

Since y_{α} is a nice point of $\Sigma_{\alpha}(C_{\alpha})$, (9) rewrites

$$C_{\alpha}(\geqslant)_{\alpha}x_{o\alpha} + \varepsilon^{\nu}\eta_{o\alpha}^{\nu} + V_{\alpha}^{*}(x) + L_{\alpha}(x\ ;\ \varepsilon^{\nu}\eta^{\nu}) + o(\varepsilon^{\nu},\eta^{\nu}) \tag{11}$$

$$\nu = 1,2,\ldots$$

and (10) rewrites

$$\Phi_{\alpha}(y_{\alpha}) + \varepsilon^{\nu}(\eta_{o\alpha}^{\nu} + L_{\alpha}(x\ ;\ \eta^{\nu}) + \frac{o(\varepsilon^{\nu},\eta^{\nu})}{\varepsilon^{\nu}}) \in \overline{B(C_{\alpha})}$$

Note that, since $y_{\alpha} \in \Sigma_{\alpha}(C_{\alpha})$, we have $\Phi_{\alpha}(y_{\alpha}) = C_{\alpha}$. As $\nu \to \infty$ $\varepsilon^{\nu} \to 0$, and

$$\eta_{o\alpha}^{\nu} + L_{\alpha}(x\ ;\ \eta^{\nu}) + \frac{o(\varepsilon^{\nu},\eta^{\nu})}{\varepsilon^{\nu}} \to \eta_{o\alpha} + L_{\alpha}(x\ ;\ \eta)$$

and accordingly

$$C_{\alpha} + \eta_{o\alpha} + L_{\alpha}(x\ ;\ \eta) = C_{\alpha} + \text{grad}\ \Phi_{\alpha}(y_{\alpha})\cdot t_{\alpha} \in T(\overline{B(C_{\alpha})}, C_{\alpha})$$

From (11), by letting $\nu \to \infty$, we deduce that

$$0(\dot{\geqslant})_{\alpha}\ \text{grad}\ \Phi_{\alpha}(y_{\alpha})\cdot t_{\alpha}$$

where $(\dot{\geqslant})_{\alpha}$ is the directional preference induced by $(\geqslant)_{\alpha}$ on $T(\overline{B(C_{\alpha})}, C_{\alpha})$.

Hence we have

Lemma 3. If $y_\alpha = (x_{o\alpha}, x)$ is a nice point of $\Sigma_\alpha(C_\alpha)$, $C_\alpha \in \Omega_\alpha$ $\alpha \in \{1,2,\dots N\}$, then

$$y_\alpha + t_\alpha \in T(\overline{B/\Sigma_\alpha(C_\alpha)}, y_\alpha) \Rightarrow$$

$$C_\alpha + \text{grad}\, \Phi_\alpha(y_\alpha)\cdot t_\alpha \in T(\overline{B(C_\alpha)}, C_\alpha), \text{ and}$$

$$0 (\dot{\geqslant})_\alpha \text{ grad}\, \Phi_\alpha(y_\alpha)\cdot t_\alpha$$

where $(\dot{\geqslant})_\alpha$ is the directional preference induced by $(\geqslant)_\alpha$ on $T(\overline{B(C_\alpha)}, C_\alpha)$.

9. Theorem 3

Let $y^i_\alpha = (x^i_{o\alpha}, x^i) \in \Sigma_\alpha(C_\alpha)$, $C_\alpha \in \Omega_\alpha$, $\alpha \in \{1,2,\dots N\}$, and consider an optimal α-path $\Gamma_\alpha(\gamma^*, y^i_\alpha, s^*)$ emanating from y^i_α. From Theorem 2 we have

$$\Gamma_\alpha(\gamma^*, y^i_\alpha, s^*) \subset \Sigma_\alpha(C_\alpha), \qquad C_\alpha = x^i_{o\alpha} + V^*_\alpha(x^i)$$

Let us define the *reachable set* $E_\alpha(y^i_\alpha)$ *from* y^i_α by

$$E_\alpha(y^i_\alpha) = \{ y_\alpha = (x_{o\alpha}, x) : \ y_\alpha \in Z_\alpha, \ \exists s \text{ and } \exists \bar{\gamma}$$

such that $sC_\alpha s^*$ and $(x^i, s)\, R\, \bar{\gamma}$, and

$$y_\alpha \in \Gamma_\alpha(\bar{\gamma}, y^i_\alpha, s)\}$$

Let y^j_α, $y^j_\alpha = (x^j_{o\alpha}, x^j) \in \Gamma_\alpha(\gamma^*, y^i_\alpha, s^*)$, be a nice point of $\Sigma_\alpha(C_\alpha)$.

If $y^j_\alpha + t_\alpha \in T(E_\alpha(y^i_\alpha), y^j_\alpha)$, then there exists an infinite sequence in Z_α, namely

$$\{t^\nu_\alpha = (\eta^\nu_{o\alpha}, \eta^\nu) : \ \nu = 1,2,\dots \text{ and } \|t^\nu_\alpha - t_\alpha\| \to 0 \text{ as } \nu \to \infty\}$$

and an infinite sequence of strictly positive numbers

$$\{ \varepsilon^\nu : \ \nu = 1,2,\dots \text{ and } \varepsilon^\nu \to 0 \text{ as } \nu \to \infty \}$$

such that

$$y^j_\alpha + \varepsilon^\nu t^\nu_\alpha \in E_\alpha(y^i_\alpha) \qquad \nu = 1,2,\dots$$

and, from Theorem 1 and the definition of $E_\alpha(y^i_\alpha)$, it follows that, for ν sufficiently large

$$y^j_\alpha + \varepsilon^\nu t^\nu_\alpha \in \overline{B/\Sigma_\alpha(C_\alpha)}$$

Hence

$$T(E_\alpha(y^i_\alpha),\ y^j_\alpha) \subseteq T(\overline{B/\Sigma_\alpha(C_\alpha)},\ y^j_\alpha)$$

Then, from Lemmas 2 and 3 we have

Theorem 3. If y^j_α is a point of an optimal α-path $\Gamma_\alpha(\gamma^*,\ y^i_\alpha,\ s^*) \subset \Sigma_\alpha(C_\alpha)$ emanating from $y^i_\alpha = (x^i_{o\alpha}, x^i) \in \Sigma_\alpha(C_\alpha)$, and if y^j_α is a nice point of $\Sigma_\alpha(C_\alpha)$, then

$$y^j_\alpha + t_\alpha \in T(E_\alpha(y^i_\alpha),\ y^j_\alpha) \Rightarrow y^j_\alpha + t_\alpha \in T(\overline{B/\Sigma_\alpha(C_\alpha)},\ y^j_\alpha)$$

$$\Rightarrow 0 (\dot{\geqslant})_\alpha \operatorname{grad} \Phi_\alpha(y^j_\alpha)\cdot t_\alpha$$

where $(\dot{\geqslant})_\alpha$ is the directional preference induced by $(\geqslant)_\alpha$ on $T(\overline{B(C_\alpha)}, C_\alpha)$.

Furthermore, if $y^j_\alpha + t^*_\alpha \in T(\Gamma_\alpha(\gamma^*,\ y^i_\alpha,\ s^*),\ y^j_\alpha)$,

then $y^j_\alpha + t^*_\alpha \in T(\Sigma_\alpha(C_\alpha),\ y^j_\alpha)$ and

$$0 = \operatorname{grad} \Phi_\alpha(y^j_\alpha)\cdot t^*_\alpha$$

The last part of Theorem 3 comes from the fact that $\Gamma_\alpha(\gamma^*,\ y^i_\alpha, s^*) \subset \Sigma_\alpha(C_\alpha)$ and from Lemma 2.

C. NECESSARY CONDITIONS OF OPTIMALITY OF A STRATEGY N-TUPLE

10. Transfer of a contingent

First let us introduce

Assumption 6. For any real strictly positive number k,
$k\omega'_\alpha (\geqslant)_\alpha k\omega''_\alpha \Longleftrightarrow \omega'_\alpha (\geqslant)_\alpha \omega''_\alpha$, $\alpha = 1,2,\ldots N$.

Then we have

Lemma 4. If $y_\alpha \in \Sigma_\alpha(C_\alpha)$, $\alpha \in \{1,2,\ldots N\}$, and $t_\alpha = t_{1\alpha} + t_{2\alpha}$, where $t_{1\alpha}$ is such that

$$y_\alpha + t_{1\alpha} \in T(\Sigma_\alpha(C_\alpha), y_\alpha), \quad \text{and}$$

$$t_{2\alpha} = (\eta_{o\alpha},0), \qquad \eta_{o\alpha} \in \Omega_\alpha, \qquad 0(\dot{\geqslant})_\alpha \eta_{o\alpha}$$

where $(\dot{\geqslant})_\alpha$ is the directional preference induced by $(\geqslant)_\alpha$ on $T(\overline{B(C_\alpha)}, C_\alpha)$; then

$$y_\alpha + t_\alpha \in T(\overline{B/\Sigma_\alpha(C_\alpha)}, y_\alpha)$$

Proof. $0(\dot{\geqslant})_\alpha \eta_{o\alpha}$ implies that there exists an infinite sequence in Ω_α, namely

$$\{\eta^\nu_{o\alpha} : \nu = 1,2,\dots \text{ and } \|\eta^\nu_{o\alpha} - \eta_{o\alpha}\| \to 0 \text{ as } \nu \to \infty\}$$

and an infinite sequence of strictly positive numbers

$$\{\varepsilon^\nu : \nu = 1,2,\dots \text{ and } \varepsilon^\nu \to 0 \text{ as } \nu \to \infty\}$$

such that

$$C_\alpha(\geqslant)_\alpha C_\alpha + \varepsilon^\nu \eta^\nu_{o\alpha} \qquad \nu = 1,2,\dots$$

Then, from Assumption 4 we have

$$0(\geqslant)_\alpha \varepsilon^\nu \eta^\nu_{o\alpha} \qquad \nu = 1,2,\dots$$

and from Assumption 6 we have

$$0(\geqslant)_\alpha \eta^\nu_{o\alpha} \qquad \nu = 1,2,\dots \tag{12}$$

Let $y_\alpha = (x_{o\alpha},x)$ and $t_{1\alpha} = (\bar{\eta}_{o\alpha}, \bar{\eta})$.

Since $y_\alpha + t_{1\alpha} \in T(\Sigma_\alpha(C_\alpha), y_\alpha)$, there exists an infinite sequence in Z_α, namely

$$\{t^\nu_{1\alpha} = (\vec{\eta}^\nu_{o\alpha}, \vec{\eta}^\nu) : \quad \nu = 1,2,\dots \quad \text{and} \quad \|t^\nu_{1\alpha} - t_{1\alpha}\| \to 0 \text{ as } \nu \to \infty \}$$

and an infinite sequence of strictly positive numbers

$$\{\vec{\varepsilon}^\nu : \quad \nu = 1,2,\dots \quad \text{and} \quad \vec{\varepsilon}^\nu \to 0 \quad \text{as} \quad \nu \to \infty \}$$

such that

$$x_{o\alpha} + \vec{\varepsilon}^\nu \vec{\eta}^\nu_{o\alpha} + V^*(x + \vec{\varepsilon}^\nu \vec{\eta}^\nu) = C_\alpha \tag{13}$$

From (12) and Assumption 6 it follows that

$$0 (\geqslant)_\alpha \, \bar{\varepsilon}^\nu \, \eta^\nu_{o\alpha} \tag{14}$$

and from (13), (14) and Assumption 4 we have

$$C_\alpha (\geqslant)_\alpha x_{o\alpha} + \bar{\varepsilon}^\nu (\bar{\eta}^\nu_{o\alpha} + \eta^\nu_{o\alpha}) + V^*(x + \bar{\varepsilon}^\nu \, \bar{\eta}^\nu) \tag{15}$$

where $(\bar{\eta}^\nu_{o\alpha} + \eta^\nu_{o\alpha}, \bar{\eta}^\nu) \to t_\alpha = t_{1\alpha} + t_{2\alpha}$ as $\nu \to \infty$.

At last, from (15) it follows that

$$y_\alpha + t_\alpha \in T(\overline{B/\Sigma_\alpha(C_\alpha)}, y_\alpha)$$

which concludes the proof of Lemma 4.

Now, let $y^i_\alpha = (x^i_{o\alpha}, x^i) \in \Sigma_\alpha(C_\alpha)$, $C_\alpha \in \Omega_\alpha$, $\alpha \in \{1,2,\ldots N\}$, and consider an optimal α-path $\Gamma_\alpha(\gamma^*, y^i_\alpha, s^*)$ emanating from y^i_α . From Theorem 2 we have

$$\Gamma_\alpha(\gamma^*, y^i_\alpha, s^*) \subset \Sigma_\alpha(C_\alpha) , \qquad C_\alpha = x^i_{o\alpha} + V^*_\alpha(x^i)$$

Let us introduce

Assumption 7. For any $y_\alpha \in \Gamma_\alpha(\gamma^*, y^i_\alpha, s^*)$, $y_\alpha = (x_{o\alpha}, x)$, there exists a non singular mapping $A(x^i, x\,; \cdot)$

$$A(x^i, x\,; \cdot) \begin{cases} G \to G \\ \eta^i \mapsto \eta = A(x^i, x\,; \eta^i) \end{cases}$$

such that

$$y_\alpha + (\eta_{o\alpha}, \eta) \in T(\Sigma_\alpha(C_\alpha), y_\alpha) \Rightarrow$$

$$y^i_\alpha + (\eta_{o\alpha}, A^{-1}(x^i, x\,; \eta)) \in T(\Sigma_\alpha(C_\alpha), y^i_\alpha)$$

where $A^{-1}(x^i, x\,; \cdot)$ is the inverse of $A(x^i, x\,; \cdot)$; that is $\eta = A(x^i, x\,; \eta^i) \iff \eta^i = A^{-1}(x^i, x\,; \eta)$.

First, let us prove

Lemma 5. If Assumption 7 is satisfied and if y_α ,

$y_\alpha \in \Gamma_\alpha(\gamma^*, y_\alpha^i, s^*) \subset \Sigma_\alpha(C_\alpha)$, is a nice point of $\Sigma_\alpha(C_\alpha)$, then

$$y_\alpha + (\eta_{o\alpha}, \eta) \in T(\overline{B/\Sigma_\alpha(C_\alpha)}, y_\alpha) \Rightarrow$$

$$y_\alpha^i + (\eta_{o\alpha}, A^{-1}(x^i, x ; \eta)) \in T(\overline{B/\Sigma_\alpha(C_\alpha)}, y_\alpha^i)$$

Proof. Let us suppose that $y_\alpha + t_\alpha \in T(\overline{B/\Sigma_\alpha(C_\alpha)}, y_\alpha)$, $t_\alpha = (\eta_{o\alpha}, \eta)$, then there exists an infinite sequence in Z_α, namely

$$\{t_\alpha^\nu = (\eta_{o\alpha}^\nu, \eta^\nu) : \nu = 1,2,\dots \text{ and } \|t_\alpha^\nu - t_\alpha\| \to 0 \text{ as } \nu \to \infty\}$$

and an infinite sequence of strictly positive numbers

$$\{\varepsilon^\nu : \nu = 1,2,\dots \text{ and } \varepsilon^\nu \to 0 \text{ as } \nu \to \infty\}$$

such that

$$y_\alpha + \varepsilon^\nu t_\alpha^\nu \in \overline{B/\Sigma_\alpha(C_\alpha)} \qquad \nu = 1,2,\dots$$

that is

$$C_\alpha (\geqslant)_\alpha\, x_{o\alpha} + \varepsilon^\nu \eta_{o\alpha}^\nu + V_\alpha^*(x + \varepsilon^\nu \eta^\nu) \qquad \nu = 1,2,\dots$$

Since y_α is a nice point of $\Sigma_\alpha(C_\alpha)$ it follows that

$$C_\alpha (\geqslant)_\alpha x_{o\alpha} + \varepsilon^\nu \eta_{o\alpha}^\nu + V_\alpha^*(x) + L_\alpha(x ; \varepsilon^\nu \eta^\nu) + o(\varepsilon^\nu, \eta^\nu) \tag{16}$$

$$\nu = 1,2,\dots$$

and $\dfrac{\|o(\varepsilon^\nu, \eta^\nu)\|}{\varepsilon^\nu} \to 0$ as $\nu \to \infty$.

Since $x_{o\alpha} + V_\alpha^*(x) = C_\alpha$, (16) rewrites

$$0 (\geqslant)_\alpha\, \varepsilon^\nu \eta_{o\alpha}^\nu + \varepsilon^\nu L_\alpha(x ; \eta^\nu) + o(\varepsilon^\nu, \eta^\nu) \tag{17}$$

and since

$$\eta_{o\alpha}^\nu + L_\alpha(x ; \eta^\nu) \to \eta_{o\alpha} + L_\alpha(x ; \eta) \text{ as } \nu \to \infty$$

it follows from (17) that

$$0 (\dot{\geqslant})_\alpha\, \eta_{o\alpha} + L_\alpha(x ; \eta) \tag{18}$$

Now let

$$t_\alpha = t_{1\alpha} + t_{2\alpha}$$

$$t_{1\alpha} = (-L_\alpha(x\ ;\ \eta_\alpha),\eta)$$

$$t_{2\alpha} = (\eta_{o\alpha} + L_\alpha(x\ ;\ \eta),0)$$

One can readily verify that

$$\text{grad}\ \Phi_\alpha(y_\alpha)\cdot t_{1\alpha} = 0$$

and hence, from Lemma 2

$$y_\alpha + t_{1\alpha} \in T(\Sigma_\alpha(C_\alpha),\ y_\alpha) \tag{19}$$

On the other hand we have

$$y^i_\alpha + (\eta_{o\alpha},\ A^{-1}(x^i,x\ ;\ \eta)) =$$

$$y^i_\alpha + (-L_\alpha(x\,;\eta),\ A^{-1}(x^i,x\ ;\ \eta)) + (\eta_{o\alpha} + L_\alpha(x\,;\eta),0) \tag{20}$$

Since

$$y_\alpha + t_{1\alpha} = y_\alpha + (-L_\alpha(x\ ;\ \eta),\eta) \in T(\Sigma_\alpha(C_\alpha),y_\alpha)$$

it follows from Assumption 7 that

$$y^i_\alpha + (-L_\alpha(x\,;\eta),\ A^{-1}(x^i,x\ ;\ \eta)) \in T(\Sigma_\alpha(C_\alpha),y_\alpha) \tag{21}$$

At last, from (18), (20), (21) and Lemma 4, we have

$$y^i_\alpha + (\eta_{o\alpha},\ A^{-1}(x^i,x\ ;\ \eta)) \in T(\overline{B/\Sigma_\alpha(C_\alpha)},y^i_\alpha)$$

which concludes the proof of Lemma 5.

11. Theorem 4

At last we have

Theorem 4. If y^j_α is a point of an optimal α-path $\Gamma_\alpha(\gamma^*,y^i_\alpha,s^*) \subset \Sigma_\alpha(C_\alpha)$ emanating from $y^i_\alpha = (x^i_{o\alpha},x^i) \in \Sigma_\alpha(C_\alpha)$, $C_\alpha \in \Omega_\alpha$, $\alpha \in \{1,2,\dots N\}$, and if y^i_α and y^j_α are nice points of $\Sigma_\alpha(C_\alpha)$, then

(a) $\quad y^j_\alpha + (\eta_{o\alpha},\eta) \in T(E_\alpha(y^i_\alpha),y^j_\alpha) \Rightarrow 0(\dot{\geqslant})_\alpha\ \eta_{o\alpha} + \lambda_\alpha(x^j\ ;\ \eta)$, and

(b) $\quad y^j_\alpha +(\eta^*_{o\alpha},\eta^*) \in T(\Gamma(\gamma^*,y^i_\alpha,s^*),y^j_\alpha) \Rightarrow 0 = \eta^*_{o\alpha} + \lambda_\alpha(x^j\ ;\eta^*)$

Furthermore, if y_α^f is the terminal point of $\Gamma_\alpha(\gamma^*, y_\alpha^i, s^*)$, that is $y_\alpha^f = \Gamma_\alpha(\gamma^*, y_\alpha^i, s^*) \cap \Theta$, then

(c) $$y_\alpha^f + (\eta_{o\alpha}^f, \eta^f) \in T(\Sigma_\alpha(C_\alpha) \cap \Theta,\ y_\alpha^f) \Rightarrow 0 = \eta_{o\alpha}^f + \lambda_\alpha(x^f ; \eta^f)$$

where, for any $x \in \gamma^*$

$$\lambda_\alpha(x ; \cdot) = L_\alpha(x^i ; A^{-1}(x^i, x ; \cdot))$$

Proof. Let $t_\alpha = (\eta_{o\alpha}, \eta)$ and $y_\alpha^j + t_\alpha \in T(E_\alpha(y_\alpha^i),\ y_\alpha^j)$. Since y_α^j is a nice point of $\Sigma_\alpha(C_\alpha)$, from Theorem 3 we have

$$y_\alpha^j + t_\alpha \in T(\overline{B/\Sigma_\alpha(C_\alpha)},\ y_\alpha^j)$$

Then from Lemma 5 it follows that

$$y_\alpha^i + (\eta_{o\alpha},\ A^{-1}(x^i, x^j ; \eta)) \in T(\overline{B/\Sigma_\alpha(C_\alpha)},\ y_\alpha^i)$$

At last from Lemma 3 we deduce that

$$0 (\dot{\geqslant})_\alpha\ \eta_{o\alpha} + L_\alpha(x^i ; A^{-1}(x^i, x^j ; \eta))$$

that is

$$0 (\dot{\geqslant})_\alpha\ \eta_{o\alpha} + \lambda_\alpha(x^j ; \eta)$$

Hence, condition (a) of Theorem 4 is proved.

Now let $y_\alpha^j + t_\alpha^* \in T(\Gamma_\alpha(\gamma^*, y_\alpha^i, s^*),\ y_\alpha^j)$, $t_\alpha^* = (\eta_{o\alpha}^*, \eta^*)$.

From Theorem 3 we have

$$y_\alpha^j + t_\alpha^* \in T(\Sigma_\alpha(C_\alpha),\ y_\alpha^j)$$

From Assumption 7 it follows that

$$y_\alpha^i + (\eta_{o\alpha}^*,\ A^{-1}(x^i, x^j ; \eta^*)) \in T(\Sigma_\alpha(C_\alpha),\ y_\alpha^i)$$

Then from Lemma 2 it follows that

$$0 = \eta^{*}_{o\alpha} + L_{\alpha}(x^{i} \; ; \; A^{-1}(x^{i},x^{j} \; ; \; \eta^{*}))$$

$$= \eta^{*}_{o\alpha} + \lambda_{\alpha}(x^{j} \; ; \; \eta^{*})$$

which proves condition (b) of Theorem 4.

At last let $y^{f}_{\alpha} + t^{f}_{\alpha} \in T(\Sigma_{\alpha}(C_{\alpha}) \cap \Theta, y^{f}_{\alpha})$, $t^{f}_{\alpha} = (\eta^{f}_{o\alpha} \, , \eta^{f})$.

From the definition of a contingent it follows that

$y^{f}_{\alpha} + t^{f}_{\alpha} \in T(\Sigma_{\alpha}(C_{\alpha}), y^{f}_{\alpha})$. Then from Assumption 7 we have

$$y^{f}_{\alpha} + (\eta^{f}_{o\alpha}, A^{-1}(x^{i},x^{f} \; ; \; \eta^{f})) \in T(\Sigma_{\alpha}(C_{\alpha}), y^{f}_{\alpha})$$

and from Lemma 2 we have

$$0 = \eta^{f}_{o\alpha} + L_{\alpha}(x^{i} \; ; \; A^{-1}(x^{i},x^{f} \; ; \; \eta^{f}))$$

$$= \eta^{f}_{o\alpha} + \lambda_{\alpha}(x^{f} \; ; \; \eta^{f})$$

which concludes the proof of Theorem 4.

References

1. Blaquière, A., Quantitative Games : Problem Statement and Examples, New Geometric Aspects in *The Theory and Application of Differential Games*, edited by J.D. Grote, D. Reidel Publishing Company, 1975

2. Athans, M., and Geering, H.P., Necessary and Sufficient Conditions for Differentuable Nonscalar-Valued Functions to Attain Extrema, *IEEE Transactions on Automatic Control*, Vol.AC-18, N°2, April 1973.

MINIMAX PARETO OPTIMAL SOLUTIONS WITH APPLICATION TO LINEAR-QUADRATIC PROBLEMS

J. Medanić

"Mihailo Pupin" Institut
Belgrade, Yugoslavia

ABSTRACT

A particular Pareto optimal solution, related to a class of minimax problems and therefore called the minimax Pareto optimal solution, is defined, analysed and applied to the study of cooperative solutions in problems described by linear dynamic systems and by vector-valued quadratic criteria. Four related topics are considered: (a) solution of the deterministic linear regulator problem with a vector-valued criterion, (b) convex approximation of the solution of the Riccati matrix differential equation, (c) solution of the stochastic linear regulator problem with a vector-valued criterion, and (d) the multiple-plant cooperative control problem.

CONTENTS

1. INTRODUCTION

Pareto optimality is the solution concept employed in vector valued optimization problems and in cooperative multiplayer games. In these notes a particular Pareto optimal solution, called the minimax Pareto optimal solution, is defined, analysed and applied to the study of cooperative solutions in control problems involving linear systems and vector-valued quadratic criteria.

The notes synthesize results described in [1,2,3,4] with corrections and clarifications ad discussed in [5,6]. Four topics are discussed: (i) solution of the deterministic linear regulator problem with a vector-valued criterion, (ii) convex approximation of the solution of the Riccati matrix differential equation, (iii) solution of the stochastic linear regulator problem with a vector-valued criterion, and (iv) the multiple-plant cooperative control problem. An essential feature of (i) is the decomposition of the space of initial conditions of the state with respect to the nature of the control. This decomposition is induced by the maximization of the associated min function over the simplex of admissible Pareto multipliers. Maximization of the min function is nontrivial due to the involved dependence of the solution of the Riccati matrix differential equation on the Pareto multipliers. The convex approximation studied in (ii) is introduced to enable the approximate maximization of the min function in the class of linear-quadratic problems. The convex approximation is then used to study the stochastic linear regulator problem with a vector-valued criterion. The multiple-plant cooperative control problem is a separate but related application of the same methodology.

2. LINEAR REGULATOR PROBLEM WITH VECTOR-VALUED CRITERION

2.1. Introduction - the Linear Regulator

Consider the class of dynamic control problems defined by the linear system

$$\dot{x} = Ax + Bu, \qquad x(t_o) = x_o \tag{1}$$

and by the collection of performance criteria

$$J\,(u,i,x_o) = \frac{1}{2}\int_{t_o}^{T} (x^T Q_i x + u^T R_i u)\,dt + \frac{1}{2}x(T)^T\, F_i x(T)\,, \qquad i \in S$$

where $x(t)$ is the n-dimensional state and $u(t)$ is the m-dimensional control at time t and A, B, Q_i, R_i and F_i are matrices of appropriate dimensions and $Q_1 \geq 0$, $F_i \geq 0$, $R_i > 0$. Moreover, the pair (A,B) is completely controllable, the Q_i may be factored into $Q_i = C_i^T C_i$ and the pairs (C_i, A) are all completely observable.

It is well known that given a scalar criterion, say $J(u,i,x_o)$, the optimal control, the associated trajectory and the minimal value of the performance criterion are given in terms of the optimal gain matrix $K_i(t)$ which is the solution of the Riccati matrix differential equation

$$\dot{K}_i + A^T K_i + K_i A - K_i S_i K_i + Q_i = 0, \qquad S_i = BR_i^{-1} B^T \tag{3}$$

satisfying the terminal condition

$$K_i(T) = F_i \tag{4}$$

The optimal control is

$$u^o = u^o(x,t) = - R_i^{-1} B^T K_i(t) x, \tag{5}$$

the associated optimal trajectory is defined by

$$\dot{x} = (A - S_i K_i) x , \qquad x(t_o) = x_o \tag{6}$$

and the minimal value of the performance criterion is

$$J(u^o, i, x_o) = \frac{1}{2} x_o^T K_i(t_o) x_o . \tag{7}$$

2.2. Vector-valued linear regulator problem

Consider the vector valued optimization problem (1), (2). The problem may be viewed as an N player cooperative control problem in which the criterion $J(u,i,x_o)$ is associated with player i and the controls of all players have been aggregated into the control u since the players have agreed to play cooperatively. The Pareto optimal strategy (see notes by Leitmann) is the usnal solution concept in vector valued optimization problems and in cooperative control problems. A particular pareto optimal strategy is pursued in these notes and it is the minimax pareto optimal strategy, or equilizer strategy, and satisfies the condition

$$\max_{i\in S} J(u^*,i,x^o) \leq \max_{i\in S} J(u,i,x_o) \tag{8}$$

for all admissible u.

To determine the solution u* satisfying (8) let

$$J_m(u,\mu,x_o) = \sum_{i=1}^{r} \mu_i J(u,i,x_o) = \sum_{i=1}^{r} \mu_i [\tfrac{1}{2} \int_{t_o}^{T} (x^T Q_i x + u^T R_i u)dt + \tfrac{1}{2} x(T)^T F_i x(T)] =$$

$$= \frac{1}{2} \int_{t_o}^{T} [x^T (\sum_{i=1}^{r} \mu_i Q_i)x + u^T (\sum_{i=1}^{r} \mu_i R_i) u]dt + \frac{1}{2} x(T)^T (\sum_{i=1}^{r} \mu_i F_i) x(T) =$$

$$= \frac{1}{2} \int_{t_o}^{T} (x^T Q x + u^T R u)dt + \frac{1}{2} x(T)^T F x(T) , \tag{9}$$

where

$$Q = \sum_{i=1}^{r} \mu_i Q_i , \qquad F = \sum_{i=1}^{r} \mu_i F_i , \qquad R = \sum_{i=1}^{r} \mu_i R_i . \tag{10}$$

Theorem 1. Let the modified max function $F_m(u,x_o)$ be defined by

$$F_m(u,x_o) = \max_{\mu \varepsilon M} J_m(u,\mu_i x_o) , \tag{11}$$

where

$$M = \{\mu : \sum_{i=1}^{r} \mu_i = 1, \mu_i \geq 0 , \qquad i = 1,\ldots,N \} \tag{12}$$

and let the max function $F(u,x_o)$ be defined by

$$F(u,x_o) \overset{\Delta}{=} \max_{i \varepsilon S} J(u,i,x_o) . \tag{13}$$

Then, for all bounded x_o and $u = \{u(t) : t\varepsilon[t_o,T]\}$,

$$F_m(u,x_o) = F(u,x_o) . \tag{14}$$

Proof. Let u be an arbitrary-bounded control and x_o an arbitrary-bounded initial condition. The control and initial condition completely specify the trajectory of the system and, therefore, also

$J(u,i,x_o)$ for arbitrary i. Let $\mu^* \varepsilon M$ maximize (11) for given u and x_o. Furthermore, let $\tilde{\mu} \varepsilon M$ be such that

$$\sum_{i=1}^{r} \tilde{\mu}_i J(u,i,x_o) = \max_{i\varepsilon S} J(u,i,x_o) \overset{\Delta}{=} F(u,x_o). \tag{15}$$

From the optimality of $\mu^* \varepsilon M$

$$F_m(u,x_o) \overset{\Delta}{=} \sum_{i=1}^{r} \mu_i^* J(u,i,x_o) \geq \sum_{i=1}^{r} \mu_i J(u,i,x_o) \overset{\Delta}{=} F(u,x_o) . \tag{16}$$

Moreover, since $F(u,x_o) \geq J(u,i,x_o)$, $i\varepsilon S$, it follows

$$F_m(u,x_o) = \sum_{i=1}^{r} \mu^* J(u,i,x_o) \geq \sum_{i=1}^{r} \mu_i^* F(u,x_o) = F(u,x_o) . \tag{17}$$

<u>Theorem 2.</u> $J_m(u,\mu,x_o)$ possesses a saddle point, i.e., there exists a pair (u^*,μ^*) satisfying

$$J_m(u^*,\mu^*,x_o) = \min_{u} \max_{\mu\varepsilon M} J_m(u,\mu,x_o) = \max_{\mu\varepsilon M} \min_{u} J_m(u,\mu,x_o) . \tag{18}$$

Proof: The control u is unconstrained, hence, its domain is closed and convex. By assumption $J_m(x_o,\mu,u)$ is convex with respect to u, the set M is a closed, bounded and convex simplex in E^N, and $J_m(x_o,\mu,u)$ is linear in μ, and therefore concave in μ. Now the proof proceeds from Bensoussan's Theorem I.1. [9]. Contrary to his assumption (I.2), $J_m(x_o,\mu,u)$ is only concave (and not strictly concave) in μ. But then M is closed and bounded. Equating all the sets $E_2^{h_2}$ of Bensoussan with M and following closely his derivation of Theorem I.1., it follows that a saddle point exists.

The saddle-point theorem is now invoked to interchange the order of the min and max operations, as follows:

$$q(x_o) = \min_u \max_{\mu\in M} J_m(u,\mu,x_o) = \max_{\mu\in M} \min_u J_m(u,\mu,x_o) = \max_{\mu\in M} G(\mu,x_o), \quad (19)$$

where $G(\mu,x_o)$ is the min function defined by

$$G(\mu,x_o) = \min_u \left[\frac{1}{2} \int_{t_o}^{T} (x^TQx + u^TRu)dt + \frac{1}{2} x(T)^T Fx(T)\right]. \quad (20)$$

<u>Corollary</u>: $G(\mu,x_o)$ is a strictly concave function over M.

Proof: Let μ_1, μ_2 be arbitrary points of M. Let, moreover,

$$G(\mu_1,x_o) = \min_u J_m(u,\mu_1,x_o) = J_m(u_1,\mu_1,x_o) < J_m(u,\mu_1,x_o), \quad u \neq u_1 \quad (21)$$

$$G(\mu_2,x_o) = \min J_m(u,\mu_2,x_o) = J_m(u_2,\mu_2,x_o) < J_m(u,\mu_2,x_o), \quad u \neq u_2$$

where the strict inequalities in (21) follow from the strict convexity of $J_m(u,\mu,x_o)$ in u. Also, let

$$G(\lambda\mu_1 + (1-\lambda)\mu_2,x_o) = \min_u J_m(u,\lambda\mu_1+(1-\lambda)\mu_2,x_o)$$

$$= J_m(u_3,\lambda\mu_1 + (1-\lambda)\mu_2,x_o) \quad (22)$$

$$= \lambda J_m(u_3,\mu_1,x_o) + (1-\lambda)J_m(u_3,\mu_2,x_o)$$

It follows from (22), and (21), for $u = u_3$, that

$$\lambda G(\mu_1,x_o) + (1-\lambda)G(\mu_2,x_o) < G(\lambda\mu_1+(1-\lambda)\mu_2,x_o) \tag{23}$$

and that, therefore, $G(\mu,x_o)$ is strictly concave for arbitrary x_o.

Note that the minimization problem defined by (1) and (20) is completely equivalent to the linear regulator problem (1) - (7).

Therefore, the minimizing control, as a function of μ, is given by

$$u^* = -R^{-1}B^TKx, \tag{24}$$

where K is the solution of the matrix Riccati equation

$$\dot{K} + A^TK + KA - KSK + Q = 0, \qquad K(T) = F, \tag{25}$$

with $S = BR^{-1}B^T$. Substituting (24) into (1) results in

$$\dot{x} = (A - SK)x, \qquad x(t_o) = x_o; \tag{26}$$

and substituting it into (20) gives

$$G(\mu,x_o) = \frac{1}{2}x_o^T K(\mu,t_o)x_o, \tag{27}$$

The minimax value $q(x_o)$ and μ^* are determined by maximizing the min function:

$$q(x_o) = G(\mu^*,x_o) = \max_{\mu\in M} \frac{1}{2}x_o^T K(\mu,t_o)x_o = \frac{1}{2}x_o^T K(\mu^*,t_o)x_o. \tag{28}$$

Substituting μ^* into (24) produces the desired minimax strategy in feedback form. It should be noted that, in general, μ^* is a function of the state x_o, and therefore so is the matrix $K(\mu^*,t_o)$. As a result, the minimax feedback strategy is a linear function of the state and a nonlinear function of the initial state. The nature of the minimax solution will now be illustrated on a simple, but nontrivial example.

2.3. Example

Consider the second-order system characterized by

$$A = \begin{Vmatrix} 0 & 1 \\ 0 & 0 \end{Vmatrix} \qquad B = \begin{Vmatrix} 0 \\ 1 \end{Vmatrix} \tag{29}$$

with $S = \{1,2\}$, where the two performance criteria are characterized by

$$\begin{aligned} Q_1 &= \begin{Vmatrix} 2 & 1 \\ 1 & 1 \end{Vmatrix}, \qquad R_1 = 2, \\ Q_2 &= \begin{Vmatrix} 1 & -1 \\ -1 & 3 \end{Vmatrix}, \qquad R_2 = 1, \end{aligned} \tag{30}$$

with $F_1 = F_2 = 0$ and an infinite optimization interval $[0,\infty)$. It follows that

$$R = \sum_{i=1}^{r} \mu_i R_i = 2\mu_1 + \mu_2 = 2 - \mu_2$$

$$Q = \sum_{i-1}^{r} \mu_i Q_i = \left\| \begin{matrix} 2\mu_1 + \mu_2 & \mu_1 - \mu_2 \\ \mu_1 - \mu_2 & \mu_1 + 3\mu_2 \end{matrix} \right\| = \left\| \begin{matrix} 2 - \mu_2 & 1 - 2\mu_2 \\ 1 - 2\mu_2 & 1 + 2\mu_2 \end{matrix} \right\|$$

with $0 \leq \mu_2 \leq 1$ where μ_1 has been eliminated through $\mu_1 + \mu_2 = 1$. Solving the quadratic matrix equation

$$KA + A^T K - KSK + Q = 0 ,$$

gives for the elements of the gain matrix K:

$$k_{11} = \surd(10 - 5\mu_2) + 2\mu_2 - 1, \quad k_{12} = 2 - \mu_2$$

$$k_{12} = \surd(10 - 5\mu_2) , \tag{31}$$

and the min function takes the form

$$G(\mu_2, x_1, x_2) = \frac{1}{2} \surd 5(x_1^2 + x_2^2) \surd(2 - \mu_2) + \frac{1}{2} x_1^2 (2\mu_2 - 1) + x_1 x_2 (2 - \mu_2) \tag{32}$$

Consider now maximization of the min function. The condition for the unconstrained maximum of the min function with respect to μ_2 takes the form

$$\partial G/\partial \mu_2 = -\surd 5(x_1^2 + x_2^2)/4 \surd(2 - \mu_2) + x_1^2 - x_1 x_2 = 0 . \tag{33}$$

If $x_1^2 - x_1 x_2 < 0$, there is no analytic maximum, $\partial G/\partial \mu_2 < 0$, and the constrained maximum is achieved for $\mu_2 = 0$. In the domain $x_1^2 - x_1 x_2 > 0$, the condition for the maximum reduces to

$$\sqrt{(2-\mu_2^*)} = 5(x_1^2 + x_2^2)/4(x_1^2 - x_1x_2). \tag{34}$$

Three different regions should be distinguished: Region I, where the unconstrained maximum satisfies $\mu_2^* \leq 0$, Region II, where the unconstrained maximum satisfies $\mu_2^* \geq 1$, and Region III, where $0 < \mu_2^* < 1$. Due to the constraint on μ_2, the contrained maximum in Region I is $\mu_2^* = 0$, and the minimax pareto optimal control is linear and coincides with the control optimal with respect to $J(u,1,x_o)$. In region II, the constrained maximum is $\mu_2^* = 1$ and the minimax pareto optimal control coincides with the control optimal with respect to $J(u,2,x_o)$. In Region III, $0 < \mu_2^* < 1$, and the constrained maximum coincides with the unconstrained maximum, and μ_2 is a nonlinear function of the initial state:

$$\mu_2^* = 2 - [\sqrt{5}(x_1^2 + x_2^2)/4(x_1^2 - x_1x_2)]^2 \tag{35}$$

and the minimax pareto optimal control is not optimal with respect to either $J(u,1.x_o)$ or $J(u,2,x_o)$ but with respect to a given linear combination of the two. To determine the boundaries of Region I note that $\mu_2^* \leq 0$ if

$$5(x_1^2 + x_2^2)/4(x_1^2 - x_1x_2) \geq 2 , \tag{36}$$

from which we obtain the condition

$$\sqrt{5}(x_2 + b_1x_1)(x_2 + b_2x_1) \geq 0, \tag{37}$$

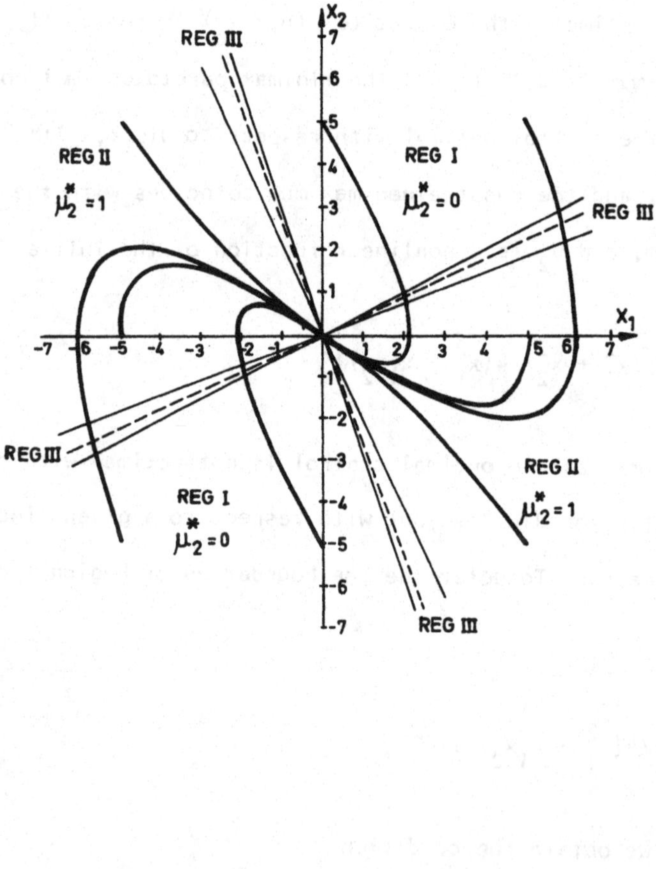

Fig. 1.

where

$$b_{1,2} = [2\sqrt{2} \pm \ (3 + 4\sqrt{10})]/\sqrt{5}. \tag{38}$$

Similarly, for Region II, note that $\mu_2^* \geq 1$ if

$$\sqrt{5}(x_1^2 + x_2^2)/4(x_1^2 - x_1x_2) \leq 1 , \tag{39}$$

from which we obtain the condition

$$\sqrt{5}(x_2 + b_3x_1)(x_2 + b_4x_1) \leq 0, \tag{40}$$

where

$$b_{3,4} = [2 \pm \sqrt{(4\sqrt{5} - 1)}]/\sqrt{5}. \tag{41}$$

The results are displayed in Fig. 1. The state space is divided by solid lines into the three regions characterizing the minimax solution, while dotted lines delineate the regions in which $J(u^*,1,x_o)$ is greater than $J(u^*,2,x_o)$ and vice versa. It may be observed that Region III, which distinguished the minimax solution, consists of those initial states where the minimal values of the two performance criteria do not differ by an amount greater than a certain bound.

2.4. Decomposition of the state space

In general, and not only in this simple example, the maximization of the min function induces a decomposition of the state space into subsets denoted by R^i, $i = 1,\dots,N$ and $R^{i_1 \dots i_s}$, $s = 2,\dots,N$. Introducing the definition of i-th vertex of M, such that

$$\mu^i = \{\mu_j : \mu_i = 1,\ \mu_j = 0, j \neq i\}, \tag{42}$$

then R^i are the closed subsets

$$R^i = \{x_o : G(x_o, \mu^i) > G(x_o, \mu),\ \mu \in M\}\ , \qquad i \in S \tag{43}$$

and $R^{i_1 \dots i_s}$ are the open subsets

$$R^{i_1 \dots i_s} = \{x_o : G(x_o, \tilde{\mu}) > G(x_o, \mu),\quad \mu \in M\ ,\ \tilde{\mu}_{i_1} > 0,\ \tilde{\mu}_{i_s} > 0,$$

$$\tilde{\mu}_j = 0,\ j \neq i_1, \dots, i_s\} \tag{44}$$

To determine the form of these subsets, let $\tilde{\mu}$ maximize for some x_o. Consider now the state $x = k\,x_o$, where k is an arbitrary scalar. Substituting into the min function gives

$$q(x) = \max_{\mu \in M} G(\mu, x) = \max_{\mu \in M} \frac{1}{2} x^T K(\mu, t)x = k^2 \max_{\mu \in M} \frac{1}{2} x_o^T K(\mu, t_o)x_o =$$

$$= k^2 \frac{1}{2} x_o^T K(\tilde{\mu}, t_o)x_o \tag{45}$$

and $\tilde{\mu}$ maximizes for x as well. Therefore, if $\tilde{\mu}$ maximizes for x_o it maximizes for all states along the line joining x_o with the origin.

By extending this line of reasoning, it follows that the state space is decomposed into subsets R^i, $i = 1,\dots,N$ and $R^{i_1\dots i_s}$, $s = 2,\dots,N$ each of which is a union of symmetric cones with a common vertex at the origin of the state space. Moreover, each separate cone is convex since its boundary is defined by a family of hyperplanes. To see this note that the condition for a minimum for $x_o \in R^{i_1\dots i_s}$ reduces to

$$x_o^T \frac{\partial K}{\partial \mu_{i_1}} x_o \triangleq x_o^T M_1 x_o = 0 \tag{46}$$

$$\vdots$$

$$x_o^T \frac{\partial K}{\partial \mu_{i_s}} x_o \triangleq x_o^T M_s x_o = 0$$

The solution of this system of equations produces $\tilde{\mu}_{i_1},\dots,\tilde{\mu}_{i_s}$ which together with $\tilde{\mu}_j = 0$, $j \neq i_1,\dots,i_s$ gives $\tilde{\mu}$. Consider now the inverse problem: For which x_o is μ maximizing? These initial states are given precisely by the x_o that satisfy (46) for $\mu = \tilde{\mu}$. Consequently, if a subregion $R^{i_1\dots i_s}$ exists (it need not in a given problem), each equation in (46) produces two hyperplanes along which μ maximizes.

2.5. Computational Aspects

Due to the linear form of the constraints defining M the gradient-projection algorithm efficiently computes the γ^* for a given x_o. For some μ_k components of the gradient are

$$g_i = \partial G/\partial\mu_i = \frac{1}{2} x_o^T[\partial K(\mu,t_o)/\partial\mu_i]x_o, \quad i = 1,\dots,N , \tag{47}$$

and the projection of the gradient on the simplex M is obtained by projecting it first on the hyperplane $\sum_{i\in S} \mu_i = 1$ with the result

$$g_\nu' = g - (e_o^T g)e_o, \tag{48}$$

where $e_o^T = (1/\sqrt{r}) \quad |1,1,\dots,1|$. It is then checked whether, for any $\mu_j = 0$, $g_{\nu j}' < 0$. If $g_{\nu j}'$ is negative, the unit direction vector e_j is appended to e_o and the projection matrix

$$E = \left\| \begin{matrix} e_o^T \\ e_1^T \\ \vdots \\ e_j^T \end{matrix} \right\| \tag{49}$$

is gradually formed where $e_k^T = |0,\dots,1,\dots,0|$, with the 1 at the kth position, whereby the final gradient projection is given by

$$g_\nu = [I - E^T(EE^T)^{-1}E]g. \tag{50}$$

The maximum is approached by the gradient algorithm

$$\mu_{k+1} = \mu_k + hg_\nu \quad , \tag{51}$$

where the stepsize h may be varied, if necessary.

Straightforward computation of the gradient demands that, beside the matrix Riccati equation, a total of r matrix equations of order n x n obtained by partial differentition of the matrix Riccati equation, and of the form

$$\partial\dot{K}/\partial\mu_i + (\partial K/\partial\mu_i)(A-SK)+(A-SK)^T(\partial K/\partial\mu_i)+KBR^{-1}R_iR^{-1}B^TK + Q_i = 0, \quad (52)$$

$$\partial K(T)/\partial\mu_i = F_i, \qquad i = 1,\dots,r ,$$

be solved. It is therefore a nontrivial simplification to show that it is possible to obtain each component of the free gradient by integrating only one scalar equation plus the state equation. Moreover, the free gradient components may in this way be obtained simultaneously, and it is therefore necessary to solve only n + r equations (plus the matrix-Riccati equation).

To determine these r auxiliary differential equations, premultiply (52) by x^T and postmultiply by x, and integrate from t_o to T to obtain

$$\int_{t_o}^{T} [x^T(\partial\dot{K}/\partial\mu_i)x + x^T(\partial K/\partial\mu_i)(A-SK)x + x^T(A-SK)^T(\partial K/\partial\mu_i)x$$

$$+ x^T KBR^{-1} R_i R^{-1} B^T Kx + x^T Q_ix]\, dt = 0. \quad (53)$$

In view of (26), it follows, after substitution into (53), that

$$\int_{t_o}^{T} [x^T(\partial\dot{K}/\partial\mu_i)x + x^T(\partial K/\partial\mu_i)\dot{x} + \dot{x}^T(\partial K/\partial\mu_i)x + u^TR_iu + x^TQ_ix]dt = 0, \quad (54)$$

or

$$\int_{t_o}^{T} (d/dt)[x^T(\partial K/\partial\mu_i)x]dt + \int_{t_o}^{T} (x^T Q_i x + u^T R_i u)\, dt = 0. \tag{55}$$

Taking into account the boundary condition of (52) and the definition (47), it follows

$$g_i = \partial G/\partial\mu_i = \frac{1}{2} x_o^T [\partial K(c,t_o)/\partial\mu_i] x_o = \frac{1}{2}\int_{t_o}^{T} (x^T Q_i x + u^T R_i u)dt + \frac{1}{2} x(T)^T F_i x(T) \tag{56}$$

or, finally,

$$g_i = J(u^*, i, x_o) . \tag{57}$$

Hence, all components of the gradient may be obtained simultaneously by integrating the state equation and a set of auxiliary equations,

$$\dot{x} = (A - SK)x , \qquad x(t_o) = x_o , \tag{58}$$

whereby

$$g_i = \frac{1}{2} y_i(T) + \frac{1}{2} x(T)^T F_i x(T) . \tag{59}$$

Note that the result expressed in (57) is identical to that which is obtained by partial differentiation of the modified criterion with respect to μ_i, assuming that the control u^* is independent of μ_i.

But this was to be expected, since we recognize that (57) is a particular case of Pagurek's result concerning the first-order variations of the performance criterion caused by variations in certain system parameters. Given a nominal μ_k, the control u* given by (24) is optimal, therefore, by Pagurek [8], the first-order variation $J_m(u^*,\mu_k,x_o)$ is the same, whether the control is implemented in open-loop or closed-loop form.

2.6. Feedback corrections

The $\mu^* = \mu^*(x_o,t_o)$ computed at t_o for x_o may need to be modified in case perturbations alter the trajectory from the optimal path defined in the solution of (26). Adhering to the goal to minimize $\max_{i\varepsilon S} J(x_o,u,i)$ the corrections are made in accordance with the following development. Let the system at time $t\varepsilon[t_o,T]$ be in the state x(t) and let u^+ be the control to be applied in the remaining time interval [t,T]. Then the problem of minimizing $\max_{i\varepsilon S} J(x_o,u,i)$ at time t is equivalent to:

$$q(x_o) = \min_{u^+} \max_{i\varepsilon S} J\,(x_o,u,i) = \min_{u^+} \max_{i\varepsilon S}\{\frac{1}{2}\int_{t_o}^{t} (x^TQ_ix + u^TR_iu)dt + J(x_o,u^+,i)\}$$

$$= \min_{u^+} \max_{i\varepsilon S} \{y_i + J(x_o,u^+,i)\} = \min_{u^+} \max_{\mu\varepsilon M} \{ \sum_{i\varepsilon S} \mu_i y_i + \sum_{i\varepsilon S} \mu_i J(x,u^+,i)\}$$

$$= \max_{\mu\varepsilon M} \min_{u^+}\{ \sum_{i\varepsilon S} \mu_i y_i + \sum_{i\varepsilon S} \mu_i\, J(x,u^+,i)\} = \max_{\mu\varepsilon M}\{ \sum_{i\varepsilon S} \mu_i y_i + \min_{u^+} J(x,u^+,\mu)\}$$

$$= \max_{\mu \in M} \{ \sum_{i \in S} \mu_i y_i + \frac{1}{2} x^T K(t,\mu) x \} \tag{60}$$

where

$$y_i = \frac{1}{2} \int_{t_o}^{t} (x^T Q_i x + u^T R_i u)\, dt \ , \tag{61}$$

and represent the costs accumulated over the elapsed interval of play, $[t_o, t)$, along the actual trajectory. If the system actually followed the optimal trajectory computed at time t_o, the value of μ determined from (60) would be the same as that computed from (28). If not, corrections may be warranted, depending whether $\mu^*(x_o) \varepsilon R^i$ or $\mu^*(x_o) \varepsilon R^{i_1 \cdots i_s}$. In the second case $\mu^*(x,y)$ will in general differ from $\mu^*(x_o)$. In the first, it need not, depending on the extent of the difference between the actually accumulated costs y_i and costs that would have been accumulated up to time t along the optimal trajectory with no disturbances. The extent to which these differences can be tolerated without causing a change in the value of μ^* are defined by the modified boundaries of R^i which are obtained by maximizing (60) over M for all $x(t)$. If the state $x(t)$ at time t does not belong to the same region as did x_o at time t_o, a change of μ^* to a new value is necessary if the minimax pareto optimal solution over $[t_o, T]$ is to be achieved. That the boundaries will change is evident from the presence of the term $\sum_{i \in S} \mu_i y_i$ which excludes the possibility that R^i and $R^{i_1 \cdots i_s}$ will remain convex cones with a common vertex at the origin of the state space. The shape of these regions at arbitrary t will be considered in the following sections.

3. CONVEX APPROXIMATIONS

3.1. Motivation

To enable thestudy of the complete fimily of Pareto optimal solutions, and in particular to simplify the maximization of the min function necessary to determine to minimax Pareto optimal control, an analysis of the dependence of the solution $K(\mu,t)$ on μ has been performed focused on the approximation of $K(\mu,t)$ over M for all $t\varepsilon[t_o,T]$. The goal is to make the dependence of $K(\mu,t)$, and therefore of $J(x_o,u^*,\mu)$ and $u^*(x,t)$ on μ explicit. The results allow a satisfactory convex approximation of $K(\mu,t)$ to be performed, which is then used in the analysis of the feedback corrections in the vector-valued linear regulator problem, as well as in the study of the stochastic linear regulator problem and of large scale systems.

3.2. Taylor series approximations

The first order Taylor series expansion $K_1^i(\mu,t)$ of $K(\mu,t)$ about a vertex point μ^i, is defined by

$$K_1^i(\mu,t) = K(\mu^i,t) + \sum_{j=1}^{r} \delta\mu_j \frac{\partial K(\mu^i,t)}{\partial\mu_j} . \tag{62}$$

The zero-th order term is $K(\mu^i,t) = K_i(t)$ where $K_i(t)$ is the solution of (25) at the vertex μ^i.

Introducing the notation $Y^{ij}(t) = \partial K(\mu^i,t)/\partial\mu_j$ the equation for $Y^{ij}(t)$ is obtained by partial differentiation of (25) with respect to μ_j at μ^i, with the result

$$dY^{ij}(t)/dt + (A-BR_i^{-1}B^TK_i)^TY^{ij}(t) + Y^{ij}(t)(A-BR_i^{-1}B^TK_i)$$

$$K_iBR_i^{-1}R_jR_i^{-1}B^TK_i + Q_j = 0\,, \qquad Y^{ij}(T) = F_j \tag{63}$$

Decomposing $Y^{ij}(t)$ into

$$Y^{ij}(t) = K_j(t) + E_{ij}(t) \tag{64}$$

where $K_j(t)$ is the solution of (25) at the vertex μ^j, we have the following proposition:

Proposition 1. The first order Taylor series expansion of $K(\mu,t)$ at the vertex μ^i for all $t\varepsilon(t_o,T)$ is given by

$$K_1^i(\mu,t) = \sum_{j=1}^{r} \mu_jY^{ij}(t) = \sum_{j=1}^{r} \mu_j[K_j(t) + E_{ij}(t)] \tag{65}$$

where $K_j(t)$ is the solution of the Riccati equation at vertex μ^j and $E_{ij}(t)$ is the solution of

$$dE_{ij}(t)/dt+(A-BR_i^{-1}B^TK_i)^TE_{ij}(t)+E_{ij}(t)(A-BR_i^{-1}B^TK_i)+Q_{ij}=0,\; E_{ij}(T)=0 \tag{66}$$

while

$$Q_{ij} = (R_i^{-1}B^TK_i - R_j^{-1}B^TK_j)^T R_j (R_i^{-1}B^TK_i - R_j^{-1}B^TK_j) \ . \tag{67}$$

Proof. Note that at μ^i $\delta\mu_i = \mu_i - 1$ and $\delta\mu_j = \mu_j$, $j \neq i$. Since $K_j(t)$ satisfies (26) and $Y^{ij}(t)$ satisfies (63), then $E_{ij}(t)$ because of (64) must satisfy (66). Moreover $E_{ii}(t) = 0$ because $Q_{ii} = 0$ and $E_{ii}(T) = 0$. Hence (62) becomes

$$K_1^i(\mu,t) = K_i(t) + \sum_{j=1}^{r} \delta\mu_j[K_j(t) + E_{ij}(t)]$$

$$= K_i(t) - K_i(t) - E_{ii}(t) + \sum_{j=1}^{r} \mu_j[K_j(t) + E_{ij}(t)]$$

$$= \sum_{j=1}^{r} \mu_j[K_j(t) + E_{ij}(t)].$$

3.3. Convex Approximations of the Riccati Matrix

A first order Taylor series expansion at μ^i is a tangent hyperplane to $K(\mu,t)$ at μ^i. As an approximation it is good in the neighborhood of μ^i, but deteriorates as μ recedes from μ^i. A way of improving the approximation is to consider higher order Taylor series expansions about a vertex point. However, this implies a preference to a certain vertex for which there is no reason. To approximate $K(\mu,t)$ evenly over M an approximation which approximates $K(\mu,t)$ by $K_1^i(\mu,t)$ in the neighborhood of vertex μ^i and then gradually transforms into $K_1^j(\mu,t)$ as μ approaches the neighborhood of μ^j is constructed.

Definition 1. The convex approximation of $K(\mu,t)$ for all $t\varepsilon(t_o,t_f)$ is given by

$$K_c(\mu,t) = (1-\alpha)\, K_1(\mu,t) + \alpha K_2(\mu,t) \tag{68}$$

where $K_1(\mu,t)$ is the convex combination of the zero-th order Taylor series expansions of $K(\mu,t)$ at the vertices of M,

$$K_1(\mu,t) = \sum_{i=1}^{r} \mu_i K(\mu^i,t) \tag{69}$$

and $K_2(\mu,t)$ is the convex combination of the first order Taylor series expansions of $K(\mu,t)$ at the vertices of M,

$$K_2(\mu,t) = \sum_{i=1}^{r} \mu_i K_1^i(\mu,t) \tag{70}$$

and α is a scalar parameter, $0 \leq \alpha \leq 1$.

We now show that K_1 and K_2 are lower and upper bound of $K(\mu,t)$:

Theorem 3. For each $\mu\varepsilon M$ and all $t\varepsilon(t_o,T)$

$$K_1(\mu,t) \leq K(\mu,t) \leq K_2(\mu,t)\ . \tag{71}$$

Proof: Introducing the notation

$$J^*(x_o,\mu) = 1/2\ x_o^T K(\mu,t_o)x_o \tag{72a}$$

$$G_1(x_o,\mu) = 1/2\ x_o^T K_1(\mu,t_o)x_o \tag{72b}$$

$$G_2(x_o,\mu) = 1/2\ x_o^T K_2(\mu,t_o)x_o \tag{72c}$$

$$G_1^i(x_o,\mu) = 1/2\ x_o^T K_1^i(\mu,t_o)x_o \tag{72d}$$

note that the (71) implies

$$G_1(x_o,\mu) \leq J^*(x_o,\mu) \leq G_2(x_o,\mu). \tag{73}$$

Note now that $J^*(x_o,\mu)$ is a concave function of μ in M for arbitrary x_o, because $J(x_o,u,\mu)$ is convex in u for $\mu\varepsilon M$ and is concave in $\mu\varepsilon M$ for all u. Thus, after minimization of $J(x_o,u,\mu)$ with respect to u, the resulting minimal value $J^*(x_o,\mu)$ is a concave function of μ in M for arbitrary x_o. Because it is concave, we have in particular

$$\sum_{i=1}^{r} \mu_i J^*(x_o,\mu^i) \leq J^*(x_o, \sum_{i=1}^{r} \mu_i\mu^i) \tag{74}$$

and because

$$\sum_{i=1}^{r} \mu_i\mu^i = \mu \tag{75}$$

we have from (69) and (72b) that $G_1(x_o,\mu) \leq J^*(x_o,\mu)$. On the other hand, it follows from the results of section 2.5, and in particular due to (56) that $G_1^i(x_o,\mu)$ is a first order Taylor series expansion of $J^*(x_o,\mu)$ at μ^i and therefore the tangent hyperplane to $J^*(x_o,\mu)$ at μ^i. Then, because of the concavity of $J^*(x_o,\mu)$, it follows that $J^*(x_o,\mu) \leq G_1^i(x_o,\mu)$. Summing over all $i\varepsilon S$ gives

$$J^*(x_o,\mu) = \sum_{i=1}^{r} \mu_i J^*(x_o,\mu) \leq \sum_{i=1}^{r} \mu_i G_1^i(x_o,\mu) = G_2(x_o,\mu)$$

which completes the proof.

For $\alpha = 0$ convex approximation $K_c(\mu,t)$ reduces to its lower bound $K_1(\mu,t)$ which is a linear approximation of $K(\mu,t)$ over M and is the hyperplane joing the points of $K(\mu,t)$ at the vertices of M. For $\alpha = 1$ convex approximation $K_c(\mu,t)$ reduces to its upper bound $K_2(\mu,t)$ which is a quadratic approximation of $K(\mu,t)$ over M. For $\alpha\ \varepsilon(0,1)$ the convex approximation $K_c(\mu,t)$ is in between lower and upper bound of $K(\mu,t)$. A reasonable choice for α is $\alpha = 0,5$ since it represents a mean approximation between the bounds and, in particular, for a second order algebric function the convex approximation (68) with $\alpha = 0,5$ is exact:

<u>Definition 2</u>. The second order convex approximation of $K(\mu,t)$ for all $t\varepsilon(t_o,T)$ is given by

$$K^*(\mu,t) = 1/2[K_1(\mu,t) + K_2(\mu,t)] \tag{76}$$

where $K_1(\mu,t)$ is given by (69) and $K_2(\mu,t)$ is given by (70). The

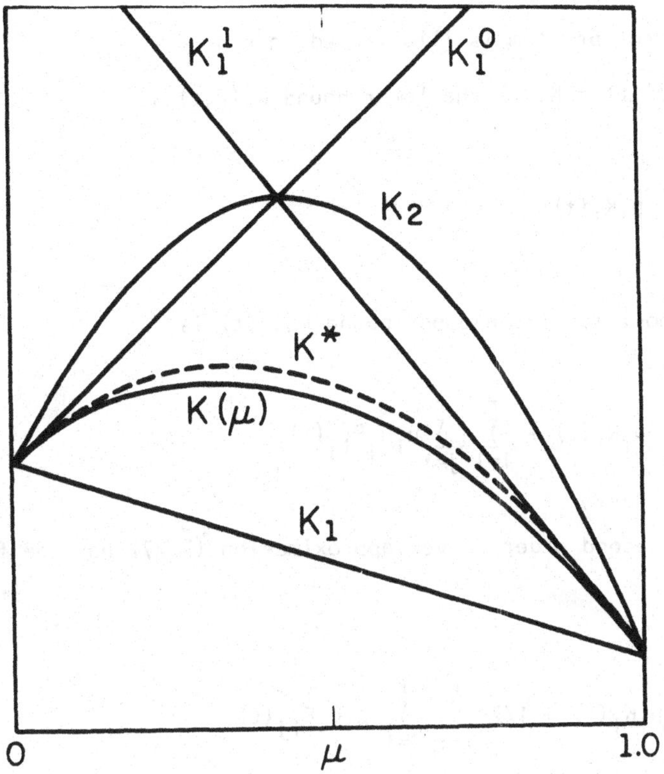

$K_1^0 = K(0) + \mu K(0)/\partial\mu$

$K_1^1 = K(1) - (1-\mu)\partial K(1)/\partial\mu$

$K_1 = (1-\mu)K(0) + \mu K(1)$

$K_2 = (1-\mu)K(0) + \mu K(1) + \mu(1-\mu)[\partial K(0)/\partial\mu - K(1)/\partial\mu]$

$K^* = (1-\mu)K(0) + \mu K(1) + 0.5\mu(1-\mu)[\partial K(0)/\partial\mu - K(1)/\partial\mu]$

Figure 2. Lower bound K_1, upper bound K_2 and second order approximation K^* of $K(\mu)$.

defined approximations for a scalar matrix K, computed for the time-infinite control problem is illustrated in Fig. 2.

Since $K(\mu^i,t) = K_i(t)$ the lower bound $K_1(\mu,t)$ is

$$K_1(\mu,t) = \sum_{i=1}^{r} \mu_i K_i(t) \tag{77}$$

and from Proposition 1 the upper bound $K_2(\mu,t)$ is

$$K_2(\mu,t) = \sum_{i=1}^{r} \mu_i K_1(t) + \sum_{i=1}^{r} \sum_{j=1}^{r} \mu_i \mu_j E_{ij}(t) \tag{78}$$

Finally, the second order convex approximation (3.27) has the following from

$$K^*(\mu,t) = \sum_{i=1}^{r} \mu_i K_i(t) + 1/2 \sum_{i=1}^{r} \sum_{j=1}^{r} \mu_i \mu_j E_{ij}(t) \ . \tag{79}$$

3.4. Example

$$A = \begin{Vmatrix} 1 & 1 \\ 0 & 2 \end{Vmatrix}, \quad B = \begin{Vmatrix} 0 \\ 1 \end{Vmatrix}, \quad Q_1 = \begin{Vmatrix} 1 & 0 \\ 0 & 4 \end{Vmatrix}$$

$$R_1 = 1, \qquad Q_2 = \begin{Vmatrix} 4 & 2 \\ 2 & 1 \end{Vmatrix},$$

$R_2 = 2$, $F_1 = F_2 = 0$ and $t \varepsilon [0,\infty)$. It follows that

$$Q(\mu) = \begin{Vmatrix} \mu_1 + 4\mu_2 & 2\mu_2 \\ 2\mu_2 & 4\mu_1 + \mu_2 \end{Vmatrix}, \quad R(\mu) = \mu_1 + 2\mu_2$$

and since $\mu_1 + \mu_2 = 1$.

$$R = 2 - \mu_1, \qquad Q = \begin{Vmatrix} 4 - 3\mu_1 & 2(1-\mu_1) \\ 2(1-\mu_1) & 1 + 3\mu_1 \end{Vmatrix}$$

where $0 \leq \mu_1 \leq 1$. The Riccati solutions at vertex points are:

$$K(0) = \begin{Vmatrix} 36.97 & 12.48 \\ 12.48 & 12.24 \end{Vmatrix}, \qquad K(1) = \begin{Vmatrix} 30.49 & 7.87 \\ 7.87 & 6.87 \end{Vmatrix}$$

The linear approximation of $K(\mu_1)$ is

$$K_1(\mu_1) = \begin{Vmatrix} 36.97 - 6.50\mu_1 & 12.48 - 4.61\mu_1 \\ 12.48 - 4.61\mu_1 & 12.24 - 5.37\mu_1 \end{Vmatrix}$$

μ_i		$K_1(\mu_1)$	$K_2(\mu_1)$	$K^*(\mu_1)$	$J_2(\mu_1)$	$J_2(\mu_1)$	$J^*(\mu_1)$
	k_{11}	36.966	36.966	36.966			
0.0	k_{12}	12.484	12.484	12.484	37.088	37.088	37.088
	k_{22}	12.242	12.242	12.242			
	k_{11}	36.386	36.318	36.365			
0.1	k_{12}	12.053	12.023	12.044	36.106	36.106	36.106
	k_{22}	11.719	11.705	11.715			
	k_{11}	35.795	35.671	35.754			
0.2	k_{12}	11.617	11.561	11.599	35.111	35.112	35.111
	k_{22}	11.193	11.168	11.185			
	k_{11}	35.191	35.023	35.133			
0.3	k_{12}	11.176	11.100	11.150	34.104	34.104	34.105
	k_{22}	10.665	10.631	10.654			
	k_{11}	34.573	34.376	34.501			
0.4	k_{12}	10.728	10.639	10.696	33.082	33.084	33.082
	k_{22}	10.135	10.094	10.120			
	k_{11}	33.940	33.728	33.859			
0.5	k_{12}	10.274	10.178	10.287	32.045	32.046	32.045
	k_{22}	9.601	9.558	9.584			
	k_{11}	33.291	33.081	33.206			
0.6	k_{12}	9.812	9.717	9.774	30.990	30.991	30.990
	k_{22}	9.064	9.021	9.046			
	k_{11}	32.624	32.433	32.543			
0.7	k_{12}	9.342	9.256	9.305	29.916	29.917	29.916
	k_{22}	8.523	8.484	8.506			
	k_{11}	31.936	31.786	31.869			
0.8	k_{12}	8.863	8.795	8.833	28.820	28.821	28.820
	k_{22}	7.978	7.941	7.964			
	k_{11}	31.225	31.138	31.185			
0.9	k_{12}	6.373	8.334	8.355	27.700	27.700	27.700
	k_{22}	7.428	7.410	7.410			
	k_{11}	30.487	30.487	30.487			
1.0	k_{12}	7.872	7.872	7.872	26.552	26.552	26.552
	k_{22}	6.873	6.873	6.873			

Table 1. Exact $K(\mu_1)$, linear and second order approximations $K_1(\mu)$ and and $K^*(\mu_1)$ and corresponding value functions, Example 1.

and the second order approximation is

$$K_2(\mu_1) = \begin{Vmatrix} 36.97 - 5.55\mu_1 - 0.93\mu_1^2 & 12.48 - 4.21\mu_1 - 0.40\mu_1^2 \\ 12.48 - 4.21\mu_1 - 0.40\mu_1^2 & 12.24 - 5.10\mu_1 - 0.18\mu_1^2 \end{Vmatrix}$$

Exact solution $K(\mu_1)$, linear and second order approximate solutions and corresponding value functions for the initial condition $x_o^T = [1 \quad 1]$ are given in Table 1 for different $\mu_1 \; \varepsilon[0,1]$. Comparing element by element of Riccati solutions we can see that for linear approximation error is always less than 1% and that for second order approximation is less than 0.4%.

3.5. Application of Convex approximations to vector valued linear regulator Problem

A direct application of convex approximations in the vector valued linear regulator problem is in the approximate maximization of the min function

$$G(\mu,x) = \frac{1}{2} x^T K(\mu,t_o)x \tag{80}$$

in the open loop case as well as in the maximization of the more general form of the min function

$$G(\mu,x,y) = \sum_{i\varepsilon S} \mu_i y_i + \frac{1}{2} x^T K(\mu,t)x \tag{81}$$

in the feedback case, when accumulated costs on the elased interval of play have to be taken into account.

Using the convex approximation for $K(\mu,t)$, and dropping the index t for notational convenience, the following approximate expression for $G(\mu,x)$ is obtained for the open-loop solution

$$G(\mu,x) = \frac{1}{2} x^T \{ \sum_{i\in S} \mu_i K_i + \frac{1}{2} \sum_{i\in S} \sum_{j\in S} \mu_i \mu_j E_{ij} \} x \tag{82}$$

$$= \frac{1}{2} \sum_{i\in S} \mu_i \, x^T K_i x + \frac{1}{4} \sum_{i\in S} \sum_{j\in S} \mu_i \mu_j \, x^T E_{ij} x$$

Introducing the vector v and the matrix M_o with elements

$$v_i = \frac{1}{2} x^T K_i x \tag{83}$$

$$(M_o)_{ij} = \frac{1}{2} x^T E_{ij} x$$

equation (82) reduces to

$$G(\mu,x) = \mu^T v + \frac{1}{2} \mu^T M_o \mu \tag{84}$$

to be maximized under the constraints

$$\mu^T e = 1 \tag{85a}$$

$$\mu_i \geq 0 \tag{85b}$$

where e is a vector with unit components. Note that (84), (85) define a quadratic programming problem which can be solved analytically or numerically by quadratic programming algorithms.

Considering now the situation the feedback case, and the min function (81) it is easily seen that there is a complete analogy except that v + y is substituted instead of v into (84). However, the final results are considerably affected by the appearence of the vector y since its components are independent of the state x. The boundaries of R^i, i = 1,...,r are no longer defined by intersections of hyperplanes; instead, they become hyper-hyperboloidal surfaces in E^n. This point will be illustrated on an example in the next section dealing with the stochastic version of the problem where the same phenomenon is exhibited.

4. STOCHASTIC LINEAR REGULATOR WITH A VECTOR CRITERION

4.1. Statement of the problem and method of solution

Consider now the stochastic linear regulator with a vector valued criterion. Both the open-loop solution and the feedback solution shall be obtained and compared with each other as well as with the deterministic solution. The nature of the stochastic solutions and their relationships to the deterministic solution will be discribed, and illustrated through the analysis of second order systems with two performance criteria by utilizing the convex approximation of the solution of the Riccati matrix differential equation.

Let $\{\Omega, \mathcal{F}_t, P\}$ be a probability space, ω a member of the sample space Ω, and $[t_o, T]$ a closed time interval. Let

$$dx = A(d)xdt + B(t)udt + C(t)dw(t),\ x(t_o) = x_o \tag{86}$$

be a linear stochastic differential equation, and

$$J(x_o,u,i) = E\{\frac{1}{2}\int_{t_o}^{T}(x^TQ_ix+u^TR_iu)dt + \frac{1}{2}x^T(T)H_ix(T)\,|x(t_o) = x_o\}, \tag{87}$$

$i = 1,2,\ldots,N$ a collection of performance criteria, where E is the operator of mathematical expectation; A, B and C are bounded (in Euclidean norm) and Lebesgue measurable matrices on $[t_o,T]$; $u(t) \equiv u(t,\omega)$ is a random m vector such that $u(t,\cdot)$ is measurable with respect to σ-algebra $_t$ of ω sets generated by the family of random vectors $\{w(t') - w(t'') : t_o \leq t' < t'' \leq t\}$, where $w(t)$ is the Wiener process in E^s, and E is the real line. It is also assumed that

$$\int_{t_o}^{T}||u(t,\omega)||^2\,dt = \int_{t_o}^{T}E\{u^T(t,\omega)u(t,\omega)\}dt < \infty\,. \tag{88}$$

In addition, $Q_i(t)$ and $R_i(t)$ are now measurable, locally bounded, symmetric matrices. In the stochastic linear regulator problem with vector valued criterion the problem is to determine a pair (u^*,i^*) such that

$$J(x_o,u^*,i^*) = \min_{u\in U}\max_{i\in S} J(x_o,u,i)$$

Applying the methodology developed for the deterministic problem made possible by the fact that analogs of Theorems 1 and 2 apply in the stochastic problem, the solution is determined by the following development:

$$q(x_o) = \min_{u\in U} \max_{i\in S} J(x_o,u,i) = \min_{u\in U} \max_{\mu\in M} J_m(x_o,u,\mu)$$

$$= \max_{\mu\in M} \min_{u\in U} J_m(x_o,u,\mu) = \max_{\mu\in M} G(x_o,\mu) \tag{89}$$

where now

$$G(x_o,\mu) = \min_{u\in U} \sum_{i=1}^{N} \mu_i E \frac{1}{2} \int_{t_o}^{T} (x^T Q_i x + u^T R_u u)dt + \frac{1}{2} x(T)^T H_i x(T) \mid x(t_o) = x_o$$

$$= \min_{u\in U} E \frac{1}{2} \int_{t_o}^{T} (x^T Qx + u^T Ru)dt + x(T)^T H_i x(T) \mid x(t_o) = x_o, \tag{90}$$

and Q, R and H are again given by (10). From the properties of Q_i, R_i and H_i, and the definition of M, it follows that R is positive definite, and Q and H are at least positive semidefinite.

4.2. Solution of the stochastic problem

Determination of an analytic expression for $G(x_o,\mu)$ is equivalent to solving the familiar stochastic linear regulator problem [10,11].

We will destinguish the open-loop and feedback solutions and derive them separately.

(a) Open-loop (OL) control $u^*(t)$ minimizing $J_m(x_o,u,\mu)$ for a given $\mu \varepsilon M$ is given by

$$u^*(t,\mu) = -R^{-1}B^TK\,\Phi(t,t_o)x_o \tag{91}$$

where K is the solution of (25) and $\Phi(t,t_o)$ is the fundamental matrix of the system

$$\dot{z} = (A - BR^{-1}B^T K)z, \qquad z(t_o) = x_o . \tag{92}$$

The corresponding minimum loss is

$$G(x_o,\mu) = \frac{1}{2}\,q(t_o) + \frac{1}{2}\,x_o^T K(t_o)x_o \quad , \tag{93}$$

where

$$q = -\,tr(C^T NC) \quad , \qquad q(T) = 0 \tag{94}$$

and N is the solution of

$$\dot{N} + A^T N + NA + Q = 0, \qquad N(T) = 0. \tag{95}$$

(b) Feedback (FB) control $\hat{u}^*$ for a given $\mu \varepsilon M$ has the form

$$\hat{u}^* = \hat{R}^{-1} B^T \hat{K} \hat{x}(t) \tag{96}$$

and the minimum loss over $[t_o, T]$ is given by

$$G(x_o, \hat{\mu}) = \frac{1}{2} p\,(t_o) + \frac{1}{2} x_o^T \hat{K}(t_o) x_o \tag{97}$$

$\hat{K}$ is again defined by (25) (where ^ denotes a quantity pertaining to the FB case only), in which $\{K,Q,R,H\}$ is replaced by $\{\hat{K},\hat{Q},\hat{R},\hat{H}\}$, with

$$Q = \sum_{i=1}^{r} \hat{\mu}_i Q_i, \qquad R = \sum_{i=1}^{r} \hat{\mu}_i R_i, \qquad H = \sum_{i=1}^{r} \hat{\mu}_i H_i \tag{98}$$

and $p(t)$ is the solution of

$$\dot{p} = -\,tr(C^T \hat{K} C), \qquad p(t) = 0. \tag{99}$$

In the feedback solution information on the system state gained during play will be utilized and it will therefore be to advantage to note that the minimum loss from some intermediate instant of time t to the terminal time T is given by (97) with $(t, \hat{x})$ replacing (t_o, x_o).

A minimizing control in the OL (FB) case depends on $\mu(\hat{\mu})$, and the minimax control is finally determined after maximization of $G(x_o, \mu)$ $(\hat{G}(x_o, \hat{\mu}))$. In the OL problem $\mu^* \varepsilon M$ is determined form

$$q(x_o) = G(x_o,\mu^*) = \max_{\mu\varepsilon V} \frac{1}{2} q(t_o) + \frac{1}{2} x_o^T K(t_o)x_o \tag{100}$$

and the minimax control is given by

$$u^*(t) = -R^{-1}(\mu^*)B^T K(\mu^*,t)\ \Phi(\mu^*,t,t_o)x_o \tag{101}$$

In the feedback solution one must take into account that due to random disturbance inputs, the trajectory of the system will be stochastic and continuous adjustment of μ^* from the initial value

$$\hat{q}(x_o) = \hat{G}(x_o,\hat{\mu}^*(x_o,t_o)) = \max_{\mu\varepsilon V} \frac{1}{2} p(t_o) + \frac{1}{2} x_o^T \hat{K}(t_o)x_o \tag{102}$$

may be warranted. Consider therefore an intermediate instant of time $t\varepsilon\ t_o,T$. Due to the control employed in the elapsed interval t_o,t and due to noise, the system has been brought to some state $\hat{x}(t)$ on the random trajectory along which definite costs have been accumulated for each particular criterion in the set S. Denoting these costs by

$$Y_i = \frac{1}{2}\int_{t_o}^{t} (\hat{x}^T Q_i x + \hat{u}^T R_i \hat{u})dt \tag{103}$$

and denoting by $\hat{u}^+$ the control to be used in the remaining interval of play and by U^+ the set of admissible controls over that interval, it follows that $\hat{\mu}^* = \hat{\mu}^*(\hat{x},t)$ is determined from the following determined from the following developement:

$$
\begin{aligned}
\hat{d}(x_o) &= \min_{\hat{u}\in U} \max_{i\in S} J(x_o,\hat{u},i) \\
&= \min_{u\in U} \max_{i\in S} [y_i + J(\hat{x},u^+,i)] \\
&= \min_{u\in U} \max_{\mu\in M} [\sum_{i=1}^{N} \mu_i y_i + \sum_{i=1}^{N} \hat{\mu}_i J(\hat{x},\hat{u}^+,i)] \\
&= \max_{\mu\in M} \min_{u^+\in U^+} [\sum_{i=1}^{N} \hat{\mu}_i y_i + \sum_{i=1}^{N} \hat{\mu}_i J(\hat{x},\hat{u}^+,i)] \\
&= \max_{\mu\in M} [\sum_{i=1}^{N} \hat{\mu}_i y_i + \min_{u^+\in U^+} \sum_{i=1}^{N} \hat{\mu}_i J(\hat{x},\hat{u},i)] \\
&= \max_{\mu\in M}[(\sum_{i=1}^{N} \hat{\mu}_i y_i + G(x,\hat{\mu})] \\
&= \max_{\mu\in M}[(\sum_{i=1}^{N} \hat{\mu}_i y_i + p(t) + \frac{1}{2} \hat{x}^T\hat{K}(t)\hat{x})]
\end{aligned}
\tag{104}
$$

The feedback form of the minimax control then takes the form

$$\hat{u}^* = -\hat{R}^{-1}(\hat{\mu}^*) B^T \hat{K}(\hat{\mu}^*,t)\hat{x} \tag{105}$$

with $\hat{\mu}^* = \hat{\mu}^*(\hat{x},y,t)$ as determined from (104).

4.4. Stochastic Versus Deterministic Problem and an Approximate Solution

In analysing dependence of μ^* on x_o in the open-loop solution and its dependence on x and y in the feedback solution it is convenient to draw a parallel with the deterministic problem.

Comparing (28) which defines $\mu^*(x_o)$ in the deterministic problem with its strochastic counterparts given by (93), (97) and (102), it is seen that $\frac{1}{2}q(t_o)$, $\frac{1}{2}p(t_o)$ and $\frac{1}{2}\sum_{i=1}^{r}\hat{\mu}_i y_i$, respectively are the terms distinguishing the stochastic from the deterministic problem when $C \to 0$, then $q(t_o)$ and $p(t)$, $t\varepsilon[t_o,T)]$ tend to zero and stochastic solutions in the limit reduce to the deterministic solution. The same is approximately true when the terms due to noise are negligibly small with respect to the quadratic term $\frac{1}{2}x_o^T K(t_o)x_o$ in the OL case, or the deterministic term $\frac{1}{2}\sum_{i=1}^{r}\hat{\mu}_i y_i + \frac{1}{2}\hat{x}^T K(t)\hat{x}$ in the FB case. Hence, for large $\|x_o\|$ the stochastic solution approaches the deterministic solution. On the other hand, when the quadratic term $\frac{1}{2}x_o^T K(t_o)\, x_o$ is much smaller then the other term, as when $\| x_o\|$ is small, or in the case of excessive noise, then the maximizing values of $\mu^*(x_o)$ ($\mu^*(x_o)$) depend solely on the characteristics of noise, and not on the initial state of the system.

The most interesting case occurs when both terms of $G(x_o,\mu)$ (similar results apply for the FB case as well) have balancing infulence on the value of $\mu^*(x_o)$. Analysis of this case in difficult because of the involved dependence of K on μ, and because the presènce of the noise term $\frac{1}{2}q(t_o)$ destroys the cheracteristic properties of the deterministic solution. To derive the basic characteristics of the stochastic solutions and to analyse the effects of noise on the dependence of μ^* on x_o, we utilize the convex approximation (79).

In should be noted that $E_{ij}(t)$ are positive semidefinite for $t\varepsilon[t_o,T]$, a property to be employed in the sequel. This follows from the stability of the matrices $A - S_i K_i$, and the positive semidefiniteness of Q_{ij}.

To apply these results note that treating (95) as a special case of (25) we have that N(t) may be written exectly as

$$N(t) = \sum_{i=1} \mu_i N_i(t) \quad , \tag{106}$$

where $N_i(t)$ is the solution of (95) with the ordered pair {Q,H} replaced by $\{Q_i,H_i\}$.

Then, with

$$\dot{L}_i = - tr(C^T N_i C) \quad , \qquad L_i(T) = 0 \tag{107}$$

we have the following approximate expression for (93):

$$G^*(x_o,\hat{\mu}) = \frac{1}{2} [\sum_{i=1}^{N} \hat{\mu}_i \hat{L}_i(t_o) + x_o^T(\sum_{i=1}^{N} \mu_i K_i(T_o) + + \frac{1}{2} \sum_{i=1}^{N} \sum_{j=1}^{N} \mu_i \mu_j E_{ij}(t_o))x_o] \tag{108}$$

Similarly, in the feedback case instead of (97) the following approximate expression is obtained:

$$G^*(x_o,\hat{\mu}) = \frac{1}{2} [\sum_{i=1}^{N} \hat{\mu}_i \hat{L}_i(t_o) + \frac{1}{2} \sum_{i=1}^{N} \sum_{j=1}^{N} \hat{\mu}_i \hat{\mu}_j M_{ij}(t_o) + +x_o^T(\sum_{i=1}^{N} \hat{\mu}_i K_i(t_o) + \frac{1}{2} \sum_{i=1}^{N} \sum_{j=1}^{N} \hat{\mu}_i \hat{\mu}_j E_{ij}(t_o))x_o], \tag{109}$$

and instead of (104) the approximate expression

$$\hat{G}^*(x_o,\hat{v}) = \frac{1}{2}[\sum_{i=1}^{N}\hat{\mu}_i y_i + \sum_{i=1}^{N}\hat{\mu}_i\hat{L}_i(t) + \frac{1}{2}\sum_{i=1}^{N}\sum_{j=1}^{N}\hat{\mu}_i\hat{\mu}_j M_{ij}(t) +$$

$$+ \hat{x}^T(\sum_{i=1}^{N}\hat{\mu}_i K_i(t) + \frac{1}{2}\sum_{i=1}^{N}\sum_{j=1}^{N}\hat{\mu}_i\hat{\mu}_j E_{ij}(t))\hat{x}], \qquad (110)$$

where now

$$\dot{L}_i = - tr(C^T K_i C) , \qquad L_i(T) = 0, \qquad (111)$$

$$\dot{M}_{ij} = - tr(C^T E_{ij} C) , \qquad M_{ij}(T) = 0. \qquad (112)$$

4.4. Analysis of Second Order Systems with two Criteria

The maximizing μ^* in both the OL and FB cases can be determined by solving a family of quadratic programming problems. However, at this point it is of interest to point out some general properties of the solutions. To this end we make a rather complete study of second order problems involving two criteria ($r = 2$, $n = 2$).

Consider first the OL solution. Now (108) takes the simple form

$$G(x_o,\mu) = \frac{1}{2}(\mu_1 L_1 + \mu_2 L_2) + \frac{1}{2}x_o^T(\mu_1 K_1 + \mu_2 K_2 + \frac{1}{2}\mu_1\mu_2(E_{12}+E_{21}))x_o$$

$$= \frac{1}{2}[\mu_1(L_1 - L_2) + L_2 + x_o^T(\mu_1(K_1 - K_2) + K_2 + \frac{\mu_1(1-\mu_2)}{2}E)x_o], \qquad (113)$$

where $L_i = L_i(t_o)$, $i = 1,2$ are given by (107), and $E = E_{12} + E_{21}$. The problem of determining maximum of $G(x_o,\mu)$ subject to the constraint $0 \leq \mu_1 \leq 1$, is in view of the concavity of $G(x_o,\mu)$, reduced to the problem of analysing the free extremum μ^f given by

$$\mu_1^f = \frac{L_1 - L_2 + x_o^T(K_1 - K_2 + \frac{1}{2} E)x_o}{\frac{1}{2} x_o^T Ex_o} \tag{114}$$

and deducing from it the constrained extremum μ^*. Consequently, the subregions R_o^1 and R_o^2 of initial conditions for which $\mu_1^f \geq 1(\mu_1^* = 1)$ and $\mu_1^f \leq 0(\mu_1^* = 0)$, respectively, are from (114) defined by:

$$R_o^1 = \{x_o : x_o^T P_1 x_o \geq -(L_1 - L_2) = d\} , \tag{115}$$

$$R_o^2 = \{x_o : x_o^T P_2 x_o \leq -(L_1 - L_2) = d \} ,$$

where

$$P_1 = K_1 - K_2 - \frac{1}{2} E,$$

$$P_2 = K_1 - K_2 + \frac{1}{2} E = P_1 + E , \tag{116}$$

$$d = - tr \int_{t_o}^{T} C^T(N_1 - N_2)Cdt .$$

Right hand sides of the inequalities defining R_o^1 and R_o^2 do not depend on x_o, and, moreover, with $C = 0$ the right hand sides reduce to zero and R_o^1 and R_o^2 reduce to R^1 and R^2 and the problem to the deterministic one. In order to facilitate the comparison of the stochastic with the deterministic solution recall the results of Example 1 in Section 2.3.

On the deterministic level there exist two possible and qualitatively different cases: (i) regions R^{12}, R^1 and R^2 all exist, (ii) state space coincides with either R^1 or R^2 (Case (ii) occurs if one criterion dominates the other for all x_o so that the minimax control is that which minimizes the dominating criterion). In analysing the problem on the stochastic level we restrict the analysis to the stochastic counterpart of case (i) since the nontrivial results pertinent to the stochastic counterpart of case (ii) are contained in it.

Consider therefore stochastic counterpart of case (i). The existence of both subregions R^1 and R^2 implies that the subsets

$$R^1 = \{x_o : x_o^T P_1 x_o \geq 0\} \quad ,$$
$$R^2 = \{x_o : x_o^T P_2 x_o \leq 0\} \tag{117}$$

are not emply, and concequently, since both R^1 and R^2 are cones with a common vertex at the origin of the state space, that both P_1 and P_2 are non-semidefinite matrices. It therefore follows that in the stochastic case the boundaries of R_o^1 and R_o^2 are defined by hyperbolic hypersurfaces described by the equalities (115).

Two distinct subcases may be distinguished. In subcase (a) assume that $d > 0$. It follows from (115) - (117) that in relation to R^1 and R^2, the subregions R_o^1 and R_o^2 have the following properties:

(1) $0 \notin R_o^1$, (2) $0 \in R_o^2$, (3) $R_o^1 \subset R^1$, (4) $R^2 \subset R_o^2$,

(5) Boundaries of R^1 and R^2 are the asympthotes of the boundaries of R_o^1 and R_o^2, respectively.

It is now simple, at least qualitatively to draw a portrait of the state space. The results are given ing Fig. 3. with the dashed line representing the partition of the state space form Fig. 1 for the corresponding deterministic problem.

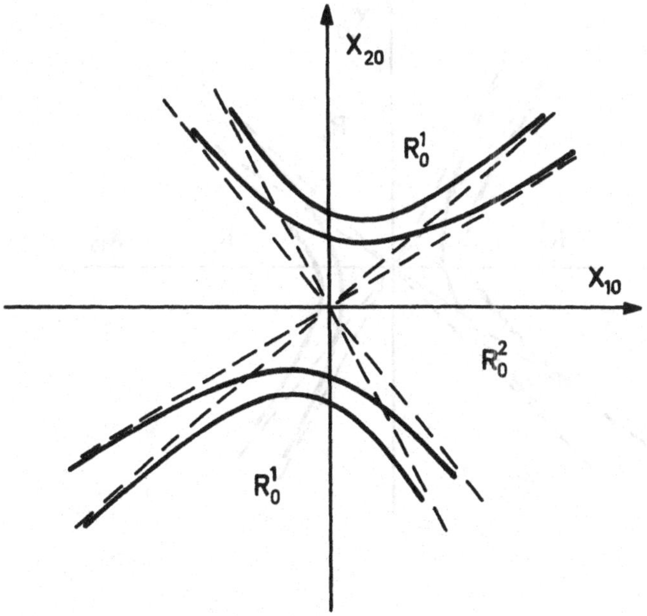

Fig. 3.

In subcase (b) assume that $d < 0$. It then follows analogously that R_o^1 and R_o^2 now have the properties:

(1) $0 \in R_o^1$, (2) $0 \notin R_o^2$, (3) $R^1 \subset R_o^1$, (4) $R_o^2 \subset R^2$,

(5) Boundaries of R^1 and R^2 are the asympthotes of the boundaries of R_o^1 and R_o^2, respectively. A qualitative illustration of this case is presented in Fig. 4.

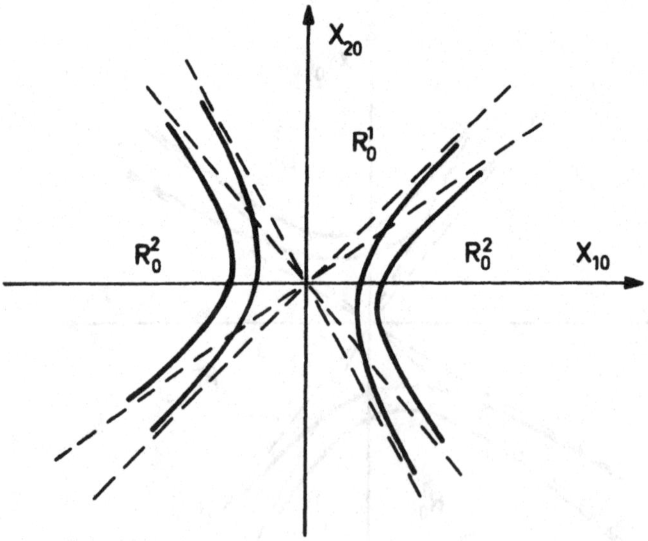

Fig. 4.

We now turn our attention to the FB solution. Consider first the situation at time $t = t_o$. The min function is given by (109) and the free extremum with respect to $\hat{\mu}_1$ is now given by

$$\hat{\mu}_1^f = \frac{\hat{L}_1 - \hat{L}_2 + \frac{1}{2} M + x_o^T(K_1 - K_2 + \frac{1}{2} E)x_o}{M + \frac{1}{2} x_o^T E\, x_o} \tag{118}$$

so that in this case $R_f^1(t_o)$ and $R_f^2(t_o)$ are given by:

$$R_f^1(t_o) = \{x_o : x_o^T P_1 x_o \geq d_1(t_o)\},$$
$$R_f^2(t_o) = \{x_o : x_o^T P_2 x_o \leq d_2(t_o)\}, \tag{119}$$

where, in view of (4.39) and (4.40), we have

$$d_1(t) = - \operatorname{tr} \int_{t_o}^{T} C^T P_1\, C dt$$
$$d_2(t_o) = - \operatorname{tr} \int_{t_o}^{T} C^T P_2\, C dt \tag{120}$$

Note that, in view of (116) and the positive semidefiniteness of E: if $d_1 < 0$ then $d_2 < d_1 < 0$, if $d_2 > 0$ then $d_1 > d_2 > 0$, so that three cases may be distinguished: (a) $d_1 > 0$ and $d_2 > 0$, (b) $d_1 < 0$ and $d_2 < 0$, (c) $d_1 > 0$ and $d_2 < 0$, while the case $d_1 < 0$ and $d_2 > 0$ is not possible.

Note that cases (a) and (b) are qualitatively equivalent to the cases (a) and (b) in the open-loop solution. Case (c) is noval and in this case $R_f^1(t_o)$ and $R_f^2(t_o)$ are found to possess the following properties:

(1) $0 \notin R_f^1(t_o)$, (2) $0 \notin R_f^1(t_o)$, (3) $R_f^1(t_o) \subset R^1$, (4) $R_f^2(t_o) \subset R^2$

(5) Boundaries of R^1 and R^2 are the asympthotes of the boundaries of $R_f^1(t_o)$ and $R_f^2(t_o)$, respectively.

Qualitative illustration of this case is presented in Fig. 5.

Finally, consider the feedback case at some arbitrary time $t \in [t_o, T]$. In this case $G(x_o, \hat{\mu})$ is given by (104) so that after analogues development we now have that $R_f^1(t)$ and $R_f^2(t)$ are given by

$$
\begin{aligned}
R_f^1(t) &= \{\hat{x} : \hat{x}^T P_1(t)\, \hat{x} \geq d_1(t) - (y_1 - y_2)\} , \\
R_f^2(t) &= \{\hat{x} : \hat{x}^T P_2(t)\, \hat{x} \leq d_2(t) - (y_1 - y_2)\} .
\end{aligned} \tag{121}
$$

As (121) shows, for arbitrary $t \in [t_o, T]$ the qualitative aspects of the solution remain the same as for t_o. It should also be noted that, in the feedback case the effect of noise and the effect of the accumulated costs in the elapsed interval of play may superimpose, or counteract. In the latter situation we may have, in a given problem, that case (a) transforms into case (b), or (c), or vice versa, depending on the particular realization of the trajectory in the elassed interval of play.

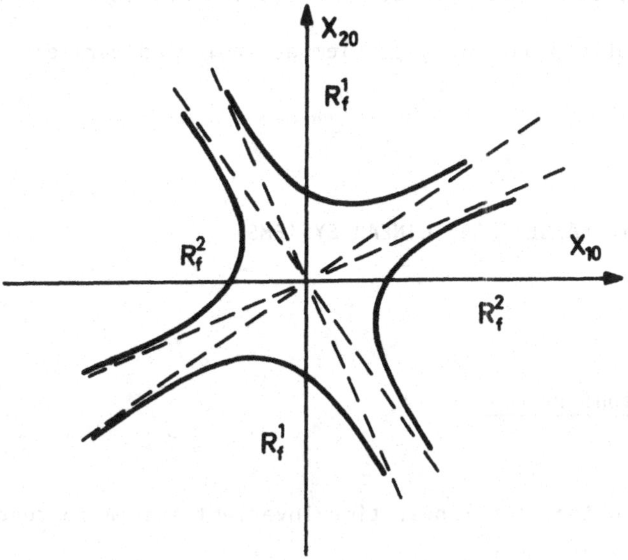

Fig. 5.

Finally, let us note that in a given problem the complete decomposition of the state space could be carried out in advance if the noise imputs are identified (matrix C given) and if the noise characteristics are assumed known. Then given the current state $\hat{x}$ and the accumulated costs y one may determine the optimal $\hat{v}^*$ from this decomposition and apply the minimax control (105).

The above results extend directly to higher dimensional problems with two criteria. In a qualitative sense the results can also be extended to multicriteria problems since the basic properties of the deterministic solution which is an asymptotic solution of the stocha" stic problem carry over to the mjlticriteria case, but the cases are to numerous to cathegorize in general. However, specific problems can be tackled with straightforward extention ot the reasoning described above. Moreover, basic features of the solution may be determined by solving a quadratic programming problem as indicated earlier.

5. APPLICATION TO LARGE SCALE LINEAR SYSTEMS

5.1. Multiple Plant Problem

Consider now that the linear time-invariant system represents N n-dimensional plants controlled by the same m-dimensional control u, that is, let in

$$dx(t)/dt = Ax(t) + Bu(t), \quad x(t_o) = x_o \tag{122}$$

x be the r-dimensional state vector, and A and B be matrices of dimension rxr and rxm. Furthermore, let the matrices A and B have the following partitioned form:

$$A = \left\| \begin{matrix} A_1 & & & 0 \\ & A_2 & & \\ & & \ddots & \\ 0 & & & A_N \end{matrix} \right\|, \quad B = \left\| \begin{matrix} B_1 \\ B_2 \\ \vdots \\ B_N \end{matrix} \right\| \tag{123}$$

For simplicity, the matrices A_i, $i = 1,\ldots,r$ are assumed to be all of order nxn so that r = nxN. This assumption simplifies notation without loss of generality.

Thus there are N subsystems S_i each characterized by the substate vector x^i such that

$$dx^i(t)/dt = A_i x^i(t) + B_i u(t), \quad x^i(t_o) = x^i_o \tag{124}$$

$$i = 1,\ldots,r$$

and

$$x^T(t) = [x^1(t), x^2(t),\ldots x^r(t)]^T .$$

Let the performance criterion of the subsystem S_i be of the form

$$J_i(x^i_o,u) = \frac{1}{2}\, x^i(t_f)^T F_i x^i(t_f) + \frac{1}{2}\int_{t_o}^{T} \{x^i(t)^T Q_i x^i(t) + u(t)^T R_i u(t)\}dt \tag{125}$$

where the matrices Q_i, R_i and F_i are of proper dimensions, Q_i and F_i being nonnegative definite, and R_i being positive definite.

If a subsystem S_i were controlled separately from other subsystems it would be possible to fulfill its objective by selecting the optimal control as a linear function of its state x^i

$$u^i(t) = -R_i^{-1} B_i^T K_i(t)\, x^i(t) \tag{126}$$

where $K_i(t)$ is the solution of the matrix Riccati differential equation

$$dK_i(t/dt + A_i^T K_i(t) + K_i(t)A_i - K_i(t)\ B_i R_i^{-1}\ B_i^T K_i(t) + Q_i = 0 \quad (127)$$

satisfying the terminal condition

$$K_i(t_f) = F_i \ .$$

However, the control $u^i(t)$ would affect other subsystems and may not be satisfactory with respect to their optimality criteria. One way to resolve this difficulty is to consider the problem as one of cooperative control and to combine the criteria $J_i(x_o^i, u)$, $i = 1,\dots,N$ into a supercriterion $J(x_o, u, \mu)$ formed by introducing a vector of weighting factors μ_i, $i = 1,\dots,r$ such that

$$J(x_o,u,\mu) = \sum_{i=1}^{N} \mu_i J_i(x_o^i,u) = \frac{1}{2}\, x(t_f)^T F(\ )x(t_f) + \frac{1}{2}\int_{t_o}^{t_f} \{x(t)^T Q(\mu)x(t) + $$

$$+ u(t)^T R(\mu)u(t)\}dt \quad (128)$$

where

$$Q(\mu) = \begin{Vmatrix} \mu_1 Q_1 & & & 0 \\ & \mu_2 Q_2 & & \\ & & \ddots & \\ 0 & & & \mu_r Q_r \end{Vmatrix}, \quad F(\mu) = \begin{Vmatrix} \mu_1 F_1 & & & 0 \\ & \mu_2 F_2 & & \\ & & \ddots & \\ 0 & & & \mu_r F_r \end{Vmatrix}, R(\mu) = \sum_{i=1}^{N} \mu_i R_i \quad (129)$$

and

$$\sum_{i=1}^{N} \mu_i = 1, \quad \mu_i \geq 0, \qquad i = 1,\dots,N.$$

The problem thus becomes a Pareto-optimal, or cooperative, control problem of generalized control theory [24].Problems of this kind are solved by defining a set of non-inferior controls which is obtained by minimizing $J(x_o,u,\mu)$ with respect to u for all $\mu\varepsilon M$. As in the vector valued linear regulator problem non-inferior controls will have the general form (24) where $K(\mu,t)$ is the solution of the matrix Riccati differential equation (25), where $Q(\mu)$, $R(\mu)$ and $F(\mu)$ are now given by (129).

Let $K(\mu,t)$ be partitioned into

$$K(\mu,t) = \begin{Vmatrix} K_{11} & K_{12} & \cdots & K_{1N} \\ K_{21} & K_{22} & \cdots & K_{2N} \\ \vdots & \vdots & & \vdots \\ K_{N1} & K_{N2} & \cdots & K_{NN} \end{Vmatrix} \tag{130}$$

where $K^T_{ij} = K^T_{ij}$ and K_{ij} are matrices of order nxn for all i and j. The dependence of K_{ij} on μ and t has been omitted in order to simplify the notation. In view of (123) and (129), the matrix Riccati differential equation in block partitioned form becomes

$$dK_{ij}/dt + A^T_iK_{ij} + K_{ij}A_j - \sum_{k=1}^{N} \sum_{\ell=1}^{N} K_{ik}B_kR^{-1}(\mu)B^T_\ell K_{\ell j} + \mu_iQ_i\delta(i-j)=0 \tag{131}$$

with the terminal condition

$$K_{ij}(\mu,t_f) = \mu_iF_i\delta(i-j) \tag{132}$$

where $\delta(i-j)$ is the Kronecker delta symbol.

5.2. Taylor series Approximations

Proposition 2. The zero-th order term $K(\mu^i,t)$ of a first order Taylor series (62) is given by

$$K(\mu^i,t) = \left\| \begin{matrix} 0 & 0 & \cdots & 0 \\ 0 & 0 & \cdots & 0 \\ \vdots & \vdots & K_{ii} & \vdots \\ 0 & 0 & \cdots & 0 \end{matrix} \right\| \qquad (133)$$

where $K_{ii}(t)$ is the solution of (127).

Prooof. The matrix Riccati differential equation (25) satisfies the necessary and sufficient conditions for the existence and uniqueness of the solution for all $t \in (t_o,T)$ [4]. It can be easily seen that $K_{ij} = 0$ is the solution of (131) for $i \neq j$ and that (131) reduces to (127) for $i = j$. Therefore $K(\mu^i,t)$ as given by (133) is the unique solution of (25), which completes the proof.

The equation for $Y^{ij}(t) = \partial K(\mu^i,t)/\partial\mu_j$ is obtained by partial differentiation of (25) with respect to μ_j

$$dY^{ij}(t)/dt + [A-BR_i^{-1}B^TK(\mu^i,t)]^TY^{ij}(t) + Y^{ij}(t)\,[A-BR_i^{-1}B^TK(\mu^i,t)]$$

$$+ K(\mu^i,t)BR_i^{-1}R_jR_i^{-1}B^TK(\mu^i,t) + \partial Q(\mu^i)/\partial\mu_j = 0, \quad Y^{ij}(T) = \partial F(\mu^i)/\partial\mu_j \quad (134)$$

$$\partial Q\,(\mu^i)/\partial\mu_j = \begin{Vmatrix} 0 & & & & 0 \\ & \ddots & & & \\ & & Q_j & & \\ & & & \ddots & \\ 0 & & & & 0 \end{Vmatrix}, \quad \partial F(\mu^i)/\partial\mu_j = \begin{Vmatrix} 0 & & & & 0 \\ & \ddots & & & \\ & & F_j & & \\ & & & \ddots & \\ 0 & & & & 0 \end{Vmatrix} \tag{135}$$

Let $Y^{ij}(t)$ be partitioned into block matrices of order nxn in the following way

$$Y^{ij}(t) = \begin{Vmatrix} Y^{ij}_{11} & Y^{ij}_{11} & \dots & Y^{ij}_{1N} \\ Y^{ij}_{21} & Y^{ij}_{22} & \dots & Y^{ij}_{2N} \\ \vdots & & & \\ Y^{ij}_{N1} & Y^{ij}_{N2} & \dots & Y^{ij}_{NN} \end{Vmatrix}. \tag{136}$$

Then in view of (132) and (133) the matrix linear differential equation (134) in partitioned form becomes

$$dY^{ij}_{k\ell}/dt + A^T_k Y^{ij}_{k\ell} + Y^{ij}_{k\ell}A_\ell - \sum_{p=1}^{N} Y^{ij}_{kp}B_pR_i^{-1}\;B_i^TK_{i\ell}\delta(\ell-i) -$$

$$-\sum_{p=1}^{N} K_{ki}B_iR_i^{-1}B_p^TY^{ij}_{p\ell}\delta(k-i) + K_{ki}B_iR_i^{-1}R_jR_i^{-1}B_i^TK_{i\ell}\delta(k-i)\delta(\ell-i) +$$

$$+ Q_k\delta(k-j)\delta(\ell-j) = 0 \tag{137}$$

with the terminal condition

$$Y_{k\ell}^{ij}(T) = F_k \delta(k-j)\delta(\ell-j) , \tag{138}$$

where $k = 1,\dots,N$ and $\ell = 1,\dots,N$.

Proposition 3. For all k and ℓ which satisfy $k \neq i,j$ or $\ell \neq i,j$

$$Y_{k\ell}^{ij}(t) = 0 \tag{139}$$

for all $t\varepsilon(t_o,T)$.

Proof. The linear matrix differential equation (134) satisfies the necessary and sufficient conditions for the existence and uniqueness of the solution for all $t\varepsilon(t_o,T)$. Because the first r equations in (137) form a closed system of equations, and because $Y_{1\ell}^{ij}(t) = 0$, $\ell = 1,\dots,N$ satisfies these N equations it is the unique solution of these equations. The same is true for all $Y_{k\ell}^{ij}$, $k = 1,\dots,N$, $k \neq i,j$, which completes the proof.

The solution of (137) thus reduces only to the case when both k and ℓ are equal to i or j. Two cases should be distinguished, first, when $j \neq i$, and second, when $j = i$. For $j \neq i$ (137) reduces to the following four linear matrix differential equations:

$$dY^{ij}_{ii}/dt + (A_i - B_i R_i^{-1} B_i^T K_{ii})^T Y^{ij}_{ii} + Y^{ij}_{ii}(A_i - B_i R_i^{-1} B_i^T K_{ii}) - Y^{ij}_{ij} B_j R_i^{-1} B_i^T K_{ii}$$

$$- K_{ii} B_i R_i^{-1} B_j^T Y^{ij}_{ji} + K_{ii} B_i R_i^{-1} B_i^T K_{ii} = 0, \quad Y^{ij}_{ii}(t_f) = 0 \qquad (140a)$$

$$dY^{ij}_{ij}/dt + (A_i - B_i R_i^{-1} B_i^T K_{ii})^T Y^{ij}_{ij} + Y^{ij}_{ij} A_j - K_{ii} B_i R_i^{-1} B_j^T Y^{ij}_{jj} = 0,$$

$$Y^{ij}_{ij}(t_f) = 0 \qquad (140b)$$

$$dY^{ij}_{ji}/dt + A_j^T Y^{ij}_{ji} + Y^{ij}_{ji}(A_i - B_i R_i^{-1} B_i^T K_{ii}) - Y^{ij}_{jj} B_j R_i^{-1} B_i^T K_{ii} = 0,$$

$$Y^{ij}_{ji}(t_f) = 0 \qquad (140c)$$

$$dY^{ij}_{jj}/dt + A_j^T Y^{ij}_{jj} + Y^{ij}_{jj} A_j + Q_j = 0, \qquad Y^{ij}_{jj}(t_f) = F_j \quad . \qquad (140d)$$

For $j = i$, we must also have $k = \ell = i$ and (137) reduces to

$$dY^{ii}_{ii}/dt + (A_i - B_i R_i^{-1} B_i^T K_{ii})^T Y^{ii}_{ii} + Y^{ii}_{ii}(A_i - B_i R_i^{-1} B_i^T K_{ii}) + K_{ii} B_i R_i^{-1} B_i^T K_{ii}$$

$$+ Q_i = 0, \qquad Y^{ii}_{ii}(t_f) = F_i \quad . \qquad (141)$$

Note that (141) is not a particular case of any of the four equations (140) for $j = i$. The solution of (141) is related to the solutions of (140) in the following way

$$Y_{ii}^{ii} = (Y_{ii}^{ij} + Y_{ij}^{ij} + Y_{ji}^{ij} + Y_{jj}^{ij})|_{j=i} \quad . \tag{142}$$

In can be easily shown that performing the necessary index substitution in (142) and adding the four matrix differential equations we obtain the differential equation (141).

Proposition 4. The first order Taylor series expansion of $K(\mu,t)$ at the vertex μ^i and for all $t\varepsilon(t_o, T)$ is given by

$$K_1^i(\mu,t) = \sum_{j=1}^{N} \mu_j Y^{ij}(t) \tag{143}$$

where $Y^{ij}(t)$ in partitioned form is given by (136) and (137) and all block matrices in (136) are equal to zero except those given as the solution of (140) for $j \neq i$ and (141) for $j = i$.

Proof: Note that at the vertex i $\delta\mu_i = \mu_i - 1$ and $\delta\mu_j = \mu_j$, $j \neq i$. Since the only nonzero submatrix $K_{ii}(t)$ of $K(\mu^i,t)$ satisfies (127) and the only nonzero submatrix $Y_{ii}^{ii}(t)$ of $Y^{ii}(t)$ satisfies (141), it can be easily shown that $Y_{ii}^{ii}(t) = K_{ii}(t)$ and therefore $Y^{ii}(t) = K(\mu^i,t)$. Hence (124) becomes

$$K_1^i(\mu,t) = K(\mu^i,t) + \sum_{j=1}^{N} \delta\mu_j Y^{ij}(t) = K(\mu^i,t) - Y^{ii}(t) +$$

$$+ \sum_{j=1}^{N} \mu_j Y^{ij}(t) = \sum_{j=1}^{N} \mu_j Y^{ij}(t) \quad . \tag{144}$$

5.3. Convex Approximations

Definition 1. The convex approximation of $K(\mu,t)$ for all $t\varepsilon(t_o,T)$ is given by

$$K_c(\mu,t) = (1-\alpha)K_1(\mu,t) + \alpha K_2(\mu,t) \tag{145}$$

where

$$K_1(\mu,t) = \sum_{i=1}^{N} \mu_i \; K(\mu^i,t) \tag{146}$$

is the convex combination of the zero-th order Taylor series expansions of $K(\mu,t)$ at the vertices of M,

$$K_2(\mu,t) = \sum_{i=1}^{N} \mu_i K_1^i(\mu,t) \tag{147}$$

is the convex combination of the first order Taylor series expansions (62) of $K(\mu,t)$ at the vertices of M, and is a scalar parameter.

Theorem 1. For each $\mu\varepsilon M$ and all $t\varepsilon(t_o,t_f)$

$$K_1(\mu,t) \leq K(\ ,t) \leq K_2(\mu,t) \ . \tag{148}$$

Definition 2. The second order approximation of $K(\mu,t)$ for all $t\ (t_o,t_f)$ is given by

$$K^*(\mu,t) = \frac{1}{2}\,[K_1(\mu,t) + K_2(\mu,t)] \tag{149}$$

In the multiple plant problem $K(\mu^i,t)$ is given by (133) where $K_{ii}(t)$ is the solution of (127) and therefore the lower dound is

$$K_1(\mu,t) = \begin{Vmatrix} \mu_1 K_{11} & & & 0 \\ & \mu_2 K_{22} & & \\ & & \ddots & \\ 0 & & & \mu_N K_{NN} \end{Vmatrix}. \tag{150}$$

Based on Proposition 4 the upper bound $K_2(\mu,t)$ is

$$K_2(\mu,t) = \sum_{i=1}^{N} \sum_{j=1}^{N} \mu_i\, \mu_j Y^{ij}(t) \tag{151}$$

where $Y^{ij}(t)$ in partitioned form (136) is the solution of (140) for $i \neq j$ and (141) for $i = j$. Finally, the second order approximation (149) is given by

$$K^*(\mu,t) = \frac{1}{2} \sum_{i=1}^{N} \mu_i K_i(t) + \frac{1}{2} \sum_{i=1}^{N} \sum_{j=1}^{N} \mu_i\, \mu_j\, Y^{ij}(t)\ . \tag{152}$$

5.4. Multiple Plant Problem Exanple

$$A_1 = \begin{Vmatrix} 1 & 2 \\ -1 & 3 \end{Vmatrix}, \quad B_1 = \begin{Vmatrix} 1 & 1 \\ -1 & 2 \end{Vmatrix}, \quad Q_1 = \begin{Vmatrix} 2 & 1 \\ 1 & 3 \end{Vmatrix},$$

$$R_1 = \begin{Vmatrix} 1 & 0 \\ 0 & 1 \end{Vmatrix}, \quad F_1 = 0$$

$$A_2 = \begin{Vmatrix} -1 & -1 \\ 3 & -5 \end{Vmatrix}, \quad B_2 = \begin{Vmatrix} 4 & 1 \\ 3 & 2 \end{Vmatrix}, \quad Q_2 = \begin{Vmatrix} 4 & 2 \\ 2 & 2 \end{Vmatrix},$$

$$R_2 = \begin{Vmatrix} 1 & 0 \\ 0 & 1 \end{Vmatrix}, \quad F_2 = 0$$

and $t \in [0,)$. It follows that

$$A = \begin{Vmatrix} 1 & 2 & 0 & 0 \\ -1 & 3 & 0 & 0 \\ 0 & 0 & -1 & -1 \\ 0 & 0 & 3 & -5 \end{Vmatrix}, \quad B = \begin{Vmatrix} 1 & 1 \\ -1 & 2 \\ 4 & 1 \\ 3 & 2 \end{Vmatrix},$$

$$Q = \begin{Vmatrix} 2\mu_1 & & & \\ \mu_1 & 3\mu_1 & & \\ 0 & 0 & 4(1-\mu_1) & 2(1-\mu_1) \\ 0 & 0 & 2(1-\mu_1) & 2(1-\mu_1) \end{Vmatrix} \qquad R = \begin{Vmatrix} 1 & 0 \\ 0 & 1 \end{Vmatrix}$$

where $0 \leq \mu_1 \leq 1$. The pair [A,B] is controllable and the pair $[C(\ _1),A]$ is: for $\mu_1 = 0$ not detectable, for μ_1 (0,1) observable and for $\mu_1 = 1$ detectable. Eigenvalues of A are: $\lambda_1 = 2 + j$, $\lambda_2 = 2 - j$, $\lambda_3 = -2$, $\lambda_4 = -4$, and for $\mu_1 = 0$ we have two nonnegative definite solutions of algebraic Riccati equation. One is optimal

$$K_o(0) = \begin{Vmatrix} 0 & 0 & 0 & 0 \\ 0 & 0 & 0 & 0 \\ 0 & 0 & 0.40 & 0.08 \\ 0 & 0 & 0.08 & 0.12 \end{Vmatrix}$$

but leaves eigenvalues $\lambda_1 = 2 + j$ and $\lambda_2 = 2-j$ unchanged. Thus the closed loop system $A - BR^{-1}B^TK_o(0)$ is unstable. The other,

$$K_s(0) = \begin{Vmatrix} 4.45 & -0.17 & -1.13 & 0.02 \\ -0.17 & 1.23 & 0.09 & -0.02 \\ -1.13 & 0.09 & 0.69 & 0.08 \\ 0.02 & -0.02 & 0.08 & 0.12 \end{Vmatrix}$$

stabilizes the closed loop system $A - BR^{-1}B^TK_s(0)$, and will be used to form the linear and the second order approximation of $K(\mu_1)$. At the other vertex point $\mu_1 = 1$ we have the unique nonnegative definite solution

$$K(1) = \begin{Vmatrix} 1.61 & 0.01 & 0 & 0 \\ 0.01 & 1.58 & 0 & 0 \\ 0 & 0 & 0 & 0 \\ 0 & 0 & 0 & 0 \end{Vmatrix}$$

The linear approximation and second order approximation are

$$K_1(\mu_1)=\left\|\begin{matrix} 4.45 - 2.84\mu_1 & -0.16 + 0.17\mu_1 & -1.13(1-\mu_1) & 0.02(1-\mu_1) \\ -0.16 + 0.17\mu_1 & 1.23 + 0.35\mu_1 & 0.09(1-\mu_1) & -0.02(1-\mu_1) \\ -1.13(1-\mu_1) & 0.09(1-\mu_1) & 0.69(1-\mu_1) & 0.08(1-\mu_1) \\ 0.02(1-\mu_1) & -0.02(1-\mu_1) & 0.08(1-\mu_1) & 0.12(1-\mu_1) \end{matrix}\right\|$$

$$K^*(\mu_1)=\left\|\begin{matrix} 4.45 + 8.86\mu_1 - 11.70\mu_1^2 & -0.16 - 1.20\mu_1 + 1.37\mu_1^2 \\ -0.16 - 1.20\mu_1 + 1.37\mu_1^2 & 1.23 + 0.60\mu_1 - 0.25\mu_1^2 \\ -1.13 - 1.96\mu_1 + 3.9\mu_1^2 & 0.09 + 0.32\mu_1 - 0.41\mu_1^2 \\ 0.02 + 0.26\mu_1 - 0.28\mu_1^2 & -0.02 + 0.02\mu_1^2 \end{matrix}\right.$$

$$\left.\begin{matrix} -13 - 1.96\mu_1 + 3.09\mu_1^2 & 0.02 + 0.26\mu_1 - 0.28\mu_1^2 \\ 0.09 + 0.32\mu_1 - 0.41\mu_1^2 & -0.02 + 0.02\mu_1^2 \\ 0.69 + 0.34\mu_1 - 1.03\mu_1^2 & 0.08 - 0.08\mu_1 \\ 0.08 - 0.08\mu_1 & 0.12 - 0.08\mu_1 - 0.04\mu_1^2 \end{matrix}\right\|$$

The value functions for exact solution $K(\mu_1)$, linear and second order approximation of $K(\mu_1)$ and initial condition $x_o^T = [1 \;\; 1 \;\; 1]$ are given in Table 2. The largest error for linear approximation is 10% and for second order approximation is 1.9%. Eigenvalues of the closed loop systems for exact $K(\mu_1)$ and second order approximation $K^*(\mu_1)$ are given in Table 3.

μ_1	J	J_1	$(\frac{J_1-J}{J})\cdot 100\%$	J^*	$(\frac{J^*-J}{J})\cdot 100\%$
0.0	2.129	2.129	0.00	2.129	0.00
0.1	2.182	2.190	0.37	2.194	0.55
0.2	2.225	2.256	1.40	2.254	1.30
0.3	2.255	2.324	3.06	2.295	1.77
0.4	2.271	2.391	5.30	2.314	1.90
0.5	2.271	2.445	7.66	2.307	1.59
0.6	2.250	2.464	9.50	2.274	1.07
0.7	2.198	2.418	10.00	2.209	0.50
0.8	2.103	2.275	8.17	2.105	0.09
0.9	1.933	2.010	3.99	1.934	0.05
1.0	1.606	1.606	0.00	1.606	0.00

Table 2. Value functions

μ_1	Closed Loop Optimal	Closed Loop Linear Approximation	Closed Loop Second Order Approximation
0.0	-12.27 - 4.42 - 2+j - 2-j	-12.27 - 4.42 - 2+j - 2-j	-12.27 - 4.42 - 2+j - 2-j
0.1	-11.65 - 4.38 -2.16+j0.78 -2.16-j0.78	-11.40 - 4.35 -2.09+j0.51 -2.09-j0.51	-12.00 - 4.35 -2.25+j1.02 -2.25-j1.02
0.2	-11.01 - 4.32 -2.31+j0.43 -2.31-j0.43	-10.52 - 4.24 - 2.94 - 1.48	-11.61 - 4.26 -2.48+j0.92 -2.48-j0.92
0.3	-10.32 - 4.23 - 3.02 - 1.95	- 9.63 - 3.83 - 3.79 - 1.17	-11.06 - 4.10 -2.72+j0.67 -2.72-j0.67
0.4	- 9.58 - 3.95 - 3.73 - 1.75	- 8.73 -3.98+j0.53 -3.98-j0.53 0.98	-10.37 -3.66+j0.35 -3.66-j0.35 - 2.47
0.5	- 8.78 -4.00+j0.47 -4.00-j0.47 - 1.64	- 7.81 -4.12+j0.72 -4.12-j0.72 - 0.86	- 9.52 -3.94+j0.69 -3.94-j0.69 - 2.10
0.6	- 7.90 -4.15+j0.64 -4.15-j0.64 - 1.57	- 6.89 -4.23+j1.87 -4.23-j1.87 - 0.80	- 8.52 -4.13+j0.83 -4.13-j0.83 - 1.87

0.7	- 6.91	- 6.00	- 7.34
	-4.28+j0.77	-4.30+j1.03	-4.28+j0.91
	-4.28-j0.77	-4.30-j1.03	-4.28-j0.91
	- 1.53	- 0.79	- 1.67
0.8	- 5.83	- 5.33	- 6.05
	-4.37+j0.94	-4.23+j1.19	-4.37+j1.03
	-4.37-j0.94	-4.23-j1.19	-4.37-j1.03
	- 1.50	- 0.87	- 1.52
	- 5.02	- 4.93	- 5.07
	-4.17+j1.13	-3.94+j1.16	-4.16+j1.16
	-4.17-j1.13	-3.94-j1.16	-4.16-j1.16
	-1.52	- 1.08	- 1.45
	- 4.45	- 4.45	- 4.45
	- 4.00	- 4.00	- 4.00
	- 2.69	- 2.69	- 2.69
	- 2.00	- 2.00	- 2.00

Table 7. Eigenvalues,

6. CONCLUSIONS

Results presented here are constrained to the linear quadratic control problems although, in principle, the method can, be extended to a broader class of problem. Numerous applications that stem from the generalization of linear regulator problems to vector valued linear regulator problems can be envisioned. Typical application is the multiple target problem /7/. The methodology may also be generalized to the study of coalition problem in multiplayer games whereby the Pareto optimality concept, and minimax pareto solution in particular, and the Nash equilibrium concept may be used to define the rules of play in the ensuing game.

REFERENCES

1. Medanić, J. and M. Andjelić, "On a Class of differential Games Without Saddle-Point Solution", JOTA, Vol. 8, No 6, Dec. 1971.

2. Medanić, J., "Mixed Strategies in Differential Games and Their Approximation", VI Annual Princeton Conf. on Information Science and Systems, Princeton, March 1972.

3. J. Medanić and Z. Simić, "Stochastic Linear Regulator with vector-valued Criterion", IFAC - IFORS Symp. on Optimization Methods - Applied Aspects, Oct. 1974, Varna.

4. M. Andjelić "On a Convex Approximation of the Riccati Equation", Report T-9, Nov. 1973, Coordinated Science Laboratory, University of Illinois.

5. Ho, Y.C., "Comment on a paper by J. Medanić and M. Andjelić", JOTA, Vol. 10, No 3, Sept. 1972.

6. J. Medanić and M. Andjelić, Authors reply, Vol. 10, No 3, Sept. 1972.

7. J. Medanić and M. Andjelić, "Minimax Solution of the Multiple - Target Problem, IEEE Transactions on Automatic Control, Vol. AC-17, No 5, October 1972.

8. B. Pagurek, "Sensitivity of the performance of optimal control systems to plant parameter variations", IEEE Trans. Automatic Control, Vol. AC-10, pp.178-180, April, 1965.

9. Bensoussan, A., in Differential Games and Related Topics, edited by H. W. Kuhn and G. P. Szegö, North-Hol and Publ. Co., 1970.

10. Wohnam, W. M., Lecture Notes on Stochastic Control, Center for Dynamical Systems, Brown University, Providence, Rhoude Island.

11. Kushner, H. J., Stochastic Stability and Control, Academic Press, 1967.

PREFERENCE OPTIMALITY AND APPLICATIONS OF PARETO-OPTIMALITY

W. Stadler

Department of Mechanical Engineering

University of California

Berkeley, California 94720

PART I
PREFERENCE OPTIMALITY

1. INTRODUCTION

To a layman one might explain the meaning of "optimization" as follows: Consider any process whatever which may be described mathematically, and whose outcome may be influenced by a set of possible decisions. Associate with the process a numerical criterion whose value depends on the decision and corresponding outcome of the process. "Optimization" then means that the decision is to be made so as to yield a maximum or a minimum numerical value of the criterion; that is, the criterion serves as a means of comparing different decisions and their outcomes.

A more general process is one where several criteria are associated with a given process. Such a situation may be interpreted as a problem for a single decision maker who wishes to "optimize" several criteria simultaneously, or as a cooperative game where each criterion represents a player, and the players then cooperate, that is act in unison as a single player, with the desire to collectively "optimize" their criteria. The problem here will be viewed in the former context. Clearly, there is no single solution concept for this type of problem, rather the choice of what is to be "optimal" depends on the situation. To be sure, many such solution concepts exist; a partial review and bibliography of such concepts may be found in Yu and Leitmann [1]. One of these concepts is preference optimality which was introduced by Stadler [2,3].

For definiteness in all of the following discussion two basic formulations for multicriteria problems will be given next. Most vector-valued criteria problems in programming and in optimal control may be formulated in this manner; the formulation is basic in that the meaning of "optimal" has been left open. As has been mentioned, there are a number of different optimality concepts when several criteria are involved; preference optimality is one such concept. In the statement of the multicriteria programming problem the following notation is employed for points $x,y \in \mathbb{R}^n$:

(i) $x \leqslant y \Rightarrow x_i \leqslant y_i \quad \forall\, i \in I = \{1,\ldots,n\}$;

(ii) $x < y \Rightarrow x_i \leqslant y_i \quad \forall\, i \in I$, and $x \neq y$;

(iii) $x \ll y \Rightarrow x_i < y_i \quad \forall\, i \in I$.

It will be apparent later that this relation is an example of a partial ordering; it is usually called the coordinatewise ordering of $\mathbb{R}^n$. The notation here differs somewhat from the usual in that inequality is emphasized ($<, \ll$) rather than equality ($\leqslant, \leqq$) as is generally done in programming. Throughout, the use of $\leqslant$, $<$ and $\ll$ in connection with vectors is to be interpreted in this light. All the concepts concerning orderings and preorderings will be made mathematically precise in Section 3. In some instances of the following the same symbol has been used to denote different quantities; this should cause no confusion, however, since their meaning should be clear from the context.

I. The Programming Problem

Let Ω(open) $\subseteq \mathbb{R}^n$ and introduce the inequality constraints

$$f(\cdot): \Omega \to \mathbb{R}^m,$$

and the equality constraints

$$h(\cdot): \Omega \to \mathbb{R}^k,$$

so that the functional constraint set is given by

$$X = \{x \in \Omega: f(x) \leqslant 0,\ h(x) = 0\}.$$

The criterion functions are

$$g_i(\cdot): X \to \mathbb{R}, \quad i = 1,\dots,N,$$

with corresponding criterion vector

$$g(\cdot) = (g_1(\cdot),\dots,g_N(\cdot)),$$

and values $g(x) \in \mathbb{R}^N$, the criteria space.

In the programming problem as in the yet to be defined control problem frequent use is made of the attainable criteria set, that is the set of numerical values of the criterion functions $g_i(\cdot)$ which may be attained as a consequence of the available decisions.

<u>Definition 1.1</u>. Attainable criteria set for Problem I. A criterion value $y \in \mathbb{R}^N$ is attainable iff there exists an $x \in X$ such that $g(x) = y$. The attainable criteria set consists of all such attainable criterion

values; it is

$Y = \{y \in \mathbb{R}^N: y = g(x),\ x \in X\}$. □

The programming problem may be stated as: Obtain the "optimal" decision(s) $x^* \in X$ for

$g(x)$ subject to $x \in X$.

II. The Control Problem

Let the state $x \in A(\text{open}) \subset \mathbb{R}^n$ be controlled by means of a control $u(\cdot)$: $[t_0, t_1]^\dagger \to U \subset \mathbb{R}^r$ in the system equations

$$\dot{x} = f(x,u) \tag{1.1}$$

with $x(t_0) \in \theta^0 \triangleq$ the initial set, and $x(t_1) \in \theta^1 \triangleq$ the terminal set, and with the independent variable t included as a state variable $x_n = t$, so that $f_n(x,u) = 1$. Furthermore, $f(\cdot)$: $A \times U \to B(\text{open}) \subset \mathbb{R}^n$, and U is the set of permissible values of $u(\cdot)$; thus U is called the control constraint set.

It is usual to confine oneself to a set of admissible controls.

Definition 1.2. Admissible controls. A control $u(\cdot)$: $[t_0, t_1] \to U$ is admissible iff

(i) $U(\text{bounded}) \subset \mathbb{R}^r$,

(ii) $u(\cdot)$ is Lebesgue measurable,

(iii) $u(\cdot)$ generates a solution $x(\cdot)$: $[t_0, t_1] \to A$ of equation (1.1) such that $x(t_0) \in \theta^0$ and $x(t_1) \in \theta^1$. □

† $[t_0, t_1]$ may be prescribed or it may be left unspecified; this question will be treated in more detail in Section 4.

The set of admissible controls is denoted by $\mathcal{F}$; it is assumed to be non-empty.

Strictly speaking, a solution of equation (1.1) is a function $s(\cdot)$ of the initial conditions, the initial value t_0 of t, and of t; for a given initial condition x^0 and an initial value t_0 it satisfies $x(t) = s(x^0, t_0; t)$. In addition, such a solution may be non-unique, without further assumptions on $f(\cdot)$. These dependencies are suppressed since they are not relevant to the present discussion.

The criterion vector

$$g(\cdot): \mathcal{F} \to \mathbb{R}^N$$

is defined in terms of the integrals

$$g_i(u(\cdot)) = \int_{t_0}^{t_1} f_{0i}(x(t),u(t))dt, \tag{1.2}$$

$i = 1,\ldots,N$, where the dependence on the initial state has again been suppressed.

The state space $\mathbb{R}^n$ is augmented by introducing a criterion response

$$\dot{y} = f^0(x,u) \quad \text{with} \quad y(t_0) = 0, \tag{1.3}$$

where $y \in \mathbb{R}^N$, the criteria space, and where $f^0 = (f_{01},\ldots,f_{0N})$.

In this context one may now define the attainable criteria set for the optimal control problem.

Definition 1.3. Attainable criteria set for Problem II. Let $u(\cdot)$ be an admissible control and $x(\cdot)$ a corresponding solution of equation (1.1). For every $t \in [t_0, t_1]$ the attainable criteria set $K(t)$ is the set of all response points $y(t) \in \mathbb{R}^N$, where $y(\cdot)$ is a solution of (1.3). In particular, the set $K(t_1)$ is the set of all $y(t_1)$. □

See Lee and Markus [4] for a more detailed discussion of the set of attainability of a system of differential equations.

One may state the multicriteria control problem as: Obtain the "optimal" control(s) $u^*(\cdot) \in \mathcal{F}$ for

$g(u(\cdot))$ subject to $\dot{x} = f(x,u)$.

Ideally, a solution concept for these problems should tie into existing theory in programming and optimal control. This goal was pursued here; that is, the theorems developed here were based on existing theorems in these fields, wherever possible.

Preference optimality is based on the concept of preferences which is well established in mathematical economics, but has had little or no application in multicriteria decision problems. A usual assumption (e.g., see Debreu [5] there is that each of N consumers has a preference preordering $\lesssim$ over his consumption set and that he is a maximizer of preference, whereas the producer is a maximizer of profit. Subject to certain wealth and equilibrium constraints, this preordering then is used to induce a partial preordering of the combined allocation space of consumers and producers with the final objective of finding a maximal element for the partial preordering. The present procedure leans on this approach. However, instead of a preordering on the consumption (or decision) set, a preordering $\lesssim$ here is introduced on the criteria space $\mathbb{R}^N$. This preordering is induced on the attainable criteria sets Y and $K(t_1)$, and preference optimality is then defined in terms of least (or greatest) elements for the preordered set $(Y, \lesssim)$ or $(K(t_1), \lesssim)$.

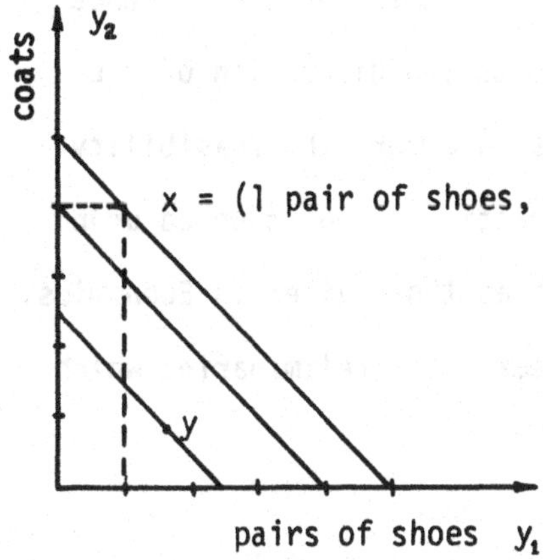

Fig. 1. A preference relation.

In a practical context the problem may be visualized in the following manner. Suppose the production of amounts $g_i(u(\cdot))$ of N different goods collectively depends on a dynamic process given by $\dot{x} = f(x,u)$, that is, corresponding to, say, a raw-material input $u(\cdot)$, the process generates amounts $g_i(u(\cdot))$, $i = 1,\ldots,N$ of the individual goods. For example, $g_1(u(\cdot))$ could be the number of pairs of leather shoes and $g_2(u(\cdot))$ the number of leather coats which are produced from a daily supply $u(\cdot)$ of raw hides. Assume now that in terms of overall efficiency the producer is indifferent among productions of batches of 8 pairs of shoes and 2 coats, or 5 pairs and 5 coats, or 7 pairs and 3 coats, and so on, as long as the total number of produced units is 10. The assumption of perfect divisibility of the goods then yields a straight line in the criteria (pairs of shoes, coats)-space; such a line is called a product transformation curve in the theory of the firm. Here it is simply called an indifference set or an equivalence class; that is, production pairs $y = (y_1, y_2)$ which lie on this line are equivalent. If one further assumes that 11 items are preferred to 10, and so on, then one obtains a preference preordering on $\mathbb{R}^2_+$ (the set $\mathbb{R}^n_+ = \{x \in \mathbb{R}^n : x \geq 0\}$), with $x \succ y$ (x preferred to y) iff $x_1 + x_2 > y_1 + y_2$ and $x \sim y$ iff $x_1 + x_2 = y_1 + y_2$. The situation is illustrated in Figure 1.

For another rudimentary discussion of the establishment of a preference relation see Samuelson [6]. Naturally a more detailed discussion of preferences will be given in subsequent sections; however, the feasibility of obtaining such a preference relation in practice is not touched upon here. The arguments pro and con are the same as those given in Economics.

The next two sections contain some mathematical preliminaries which may not be readily available.

2. MATHEMATICAL PRELIMINARIES: CONVEXITY

As one might expect, convexity plays a large part in the derivation of sufficient conditions. For the convenience of the reader some relevant theorems are quoted here without proof, with most of them stemming from Mangasarian [7]. In the cited theorems and definitions $M \subseteq \mathbb{R}^n$ and $\phi(\cdot): M \to \mathbb{R}$.

Definition 2.1. Convexity. The function $\phi(\cdot)$ is convex with respect to M at $\bar{x} \in M$ iff $x \in M$, $\theta \in [0,1]$, and $(1-\theta)\bar{x} + \theta x \in M$ imply

$$(1-\theta)\phi(\bar{x}) + \theta\phi(x) \geq \phi((1-\theta)\bar{x} + \theta x). \quad \square$$

Definition 2.2. Quasiconvexity. The function $\phi(\cdot)$ is quasiconvex with respect to M at $\bar{x} \in M$ iff $x \in M$, $\phi(x) \leq \phi(\bar{x})$, $\theta \in [0,1]$, and $(1-\theta)\bar{x} + \theta x \in M$ imply

$$\phi((1-\theta)\bar{x} + \theta x) \leq \phi(\bar{x}). \quad \square$$

Definition 2.3. Pseudoconvexity. Let $\phi(\cdot)$ be differentiable. The function $\phi(\cdot)$ is pseudoconvex with respect to M at $\bar{x} \in M$ iff $x \in M$, and $\nabla\phi(\bar{x})(x-\bar{x}) \geq 0$ imply $\phi(\bar{x}) \leq \phi(x)$. $\square$

It follows that $\phi(\cdot)$ is convex or quasiconvex on a convex set M iff $\phi(\cdot)$ is convex or quasiconvex at every point of M.

Theorem 2.1. Let $g(\cdot): M \to \mathbb{R}^N$. Assume that $g(\cdot)$ (that is, each of the component functions $g_i(\cdot)$) is convex at $\bar{x} \in M$ (convex on a convex set M), and that $c \geqslant 0$, $c \in \mathbb{R}^N$, then

$$\phi(\cdot) = cg(\cdot)$$

is convex at $\bar{x}$ (convex on a convex set M). □

Note that there is no similar theorem for quasiconvex or pseudoconvex functions.

Theorem 2.2. Let M be open. Assume that $\phi(\cdot)$ is differentiable at $\bar{x} \in M$. If $\phi(\cdot)$ is convex at $\bar{x} \in M$, then

$$\phi(x) - \phi(\bar{x}) \geqslant \nabla\phi(\bar{x})(x - \bar{x})$$

for each $x \in M$. □

Theorem 2.3. Let M be open. Let x^1, $x^2 \in M$ and assume that $\phi(\cdot)$ is differentiable and quasiconvex at x^1. Then

$$\phi(x^2) \leqslant \phi(x^1) \Rightarrow \nabla\phi(x^1)(x^2 - x^1) \leqslant 0. \quad \square$$

Theorem 2.4. Let M be convex. A necessary but not sufficient condition for $\phi(\cdot)$ to be convex on M is that the set

$$\Lambda_\nu = \{x \in M: \phi(x) \leqslant \nu\} \subset M \subset \mathbb{R}^n$$

be a convex set for each $\nu \in \mathbb{R}$. □

Theorem 2.5. Let M be convex. Let $\Lambda_\nu = \{x \in M: \phi(x) \geqslant \nu\}$. Then $\phi(\cdot)$ is quasiconcave on M iff Λ_ν is convex for each $\nu \in \mathbb{R}$. □

Theorem 2.6. Let M be open and convex and let $\phi(\cdot)$ be twice differentiable. Then $\phi(\cdot)$ is convex on M iff the quadratic form

$$\sum_{i,j=1}^{n} \phi_{ij}(x)\xi_i\xi_j\ , \quad \phi_{ij} = \frac{\partial^2\phi}{\partial x_i \partial x_j}(x)$$

is positive semidefinite in the variables $\xi_1,\ldots,\xi_n$ for every $x \in M$.

Theorem 2.7. Let $Q(\xi,\xi) = \langle \xi, Q\xi \rangle$ be a quadratic form defined on $\mathbb{R}^n$, and let Q be the corresponding n×n matrix. Let $\alpha_1,\ldots,\alpha_n$ be the eigenvalues of Q. Then the quadratic form is negative definite iff

$\alpha_i < 0$ for all $i \in \{1,\ldots,n\}$. □

3. MATHEMATICAL PRELIMINARIES: ORDERINGS AND PREFERENCES

The fundamental concept involved in all of the following discussion is that of a preference relation. In its most general form a preference relation on a set M is nothing but a binary relation on M, which has been put to a particular use, namely it is to provide a hierarchy among the elements of M. Thus, the next definition serves more as an introduction of terminology than as a definition of specific properties of such relations. By convention this definition is stated in terms of a maximizer over preferences.

Definition 3.1. Preference relation. Let M be an arbitrary set and let $x,y \in M$. A strict preference on M is a binary relation on M denoted by $\prec$. One reads "$x \prec y$" as "x is less preferred than y" or "y is preferred to x". The absence of strict preference is defined by indifference $\sim$, and one reads "$x \sim y$" as "x is indifferent to y" (or "x is equivalent to y", if $\sim$ is an equivalence relation). The juxtaposition of these two binary relations on M is a preference-indifference relation, or simply a preference relation. Thus, "$x \precsim y$" means that either $\prec$ or $\sim$ may hold. In this context one defines $\succsim$ by $y \succsim x$ iff $x \precsim y$, "$x \precsim y$ and $y \precsim x$" is written as "$x \sim y$", and "$x \precsim y$ and not $y \precsim x$" is written as "$x \prec y$" (or "$y \succ x$").† □

To make such relations a useful concept, various additional structure is imposed upon them. Among these possible additional properties, some are in such frequent use that special names have been given to preference relations which have them. The symbol " = " is to be taken as a mathematical primitive denoting "is identical with".

<u>Definition 3.2</u>. Ordering relations. Consider the following properties for a preference relation $\precsim$ defined on a set M with $x,y,z \in M$:

(i) $x \precsim x$ for every $x \in M$ (reflexivity);

(ii) "$x \precsim y$ and $y \precsim z$" $\Rightarrow$ "$x \precsim z$" (transitivity);

(iii) "$x \precsim y$ and $y \precsim x$" $\Rightarrow$ "$x = y$" (antisymmetry);

(iv) for any $x,y \in M$, either $x \precsim y$ or $y \precsim x$ or both (connexity).

Collectively one then defines:

(1) (i) and (ii) together as a partial preordering;††

(2) (i), (ii) and (iii) together as a partial ordering;

(3) the inclusion of (iv) in (1) [respectively (2)] as a complete preordering [respectively complete ordering]. □

Naturally, this by no means exhausts the properties which preference relations may have. An extensive mathematical treatment of the properties of preference relations and some of their consequences is given by

---

† The introduction of the notation $\precsim$ for a preordering is ascribed to I.N. Herstein and J. Milnor: An Axiomatic Approach to Measurable Utility, *Econometrica*, 21, 1953, pp. 291-297.

†† The term "preordering" stems from N. Bourbaki: Eléments de mathématique, Paris, Hermann, 1939- .

Fishburn [8]. Only those additional properties which are relevant to the present treatment will be defined here. In particular, the necessary conditions for preference optimality depend strongly on differentiability properties; to define these, the concept of a hypersurface in $\mathbb{R}^n$ is needed. Furthermore, the proof of Debreu's Theorem [9] leans heavily on the theory of differentiable manifolds. A good treatment of the latter subject may be found in Matsushima [10] and in Spivak [11].

Definition 3.3. Hypersurface. A set of points M in $\mathbb{R}^n$ is a C^k-hypersurface iff for every $z \in M$, there are an open neighborhood V of z, a C^k-diffeomorphism $h(\cdot)$ of V onto an open set W in $\mathbb{R}^n$ and a hyperplane H in $\mathbb{R}^n$ such that $M \cap V$ is carried by $h(\cdot)$ into $H \cap W$. □

In other words, a subset M of $\mathbb{R}^n$ is a C^k-hypersurface in $\mathbb{R}^n$ if it is an (n-1)-dimensional embedded C^k-submanifold of $\mathbb{R}^n$, where the inclusion mapping $i(\cdot): M \to \mathbb{R}^n$ may be used as the imbedding.

In defining further properties of preference relations, there are some sets which recur with enough frequency to warrant citing them here collectively. Again, let M be an arbitrary set and let $\precsim$ be a preference relation on M. For every $y \in M$, let

$$S(y) = \{x \in M: x \sim y\}$$

$$S^+(y) = \{x \in M: x \succ y\}$$

$$S^-(y) = \{x \in M: x \prec y\},$$

with

$$\bar{S}^+(y) = \{x \in M: x \succsim y\},$$

$$\bar{S}^-(y) = \{x \in M: x \precsim y\}.$$

Let their corresponding graphs in M×M be

$S = \{(x,y) \in M\times M: x \sim y\}$,

$S^+ = \{(x,y) \in M\times M: x \succ y\}$,

$S^- = \{(x,y) \in M\times M: x \prec y\}$,

$\bar{S}^+ = \{(x,y) \in M\times M: x \succsim y\}$,

$\bar{S}^- = \{(x,y) \in M\times M: x \precsim y\}$.

Properly, one should take these graphs as definitive of the corresponding preference relations.

Definition 3.4. Properties of preference relations. M is an arbitrary set unless otherwise mentioned.

(i) Continuity. A preference relation $\precsim$ on M is continuous iff for every $y \in M$ the sets $\bar{S}^+(y)$ and $\bar{S}^-(y)$ are closed in M.

(ii) Monotonicity. Let $M \subseteq \mathbb{R}^n$. A strict preference $\prec$ on M is

(1) weakly monotone iff for $x,y \in M$, $x \ll y \Rightarrow x \prec y$;

(2) monotone iff for $x,y \in M$, $x < y \Rightarrow x \prec y$.

(iii) Differentiability. A preference relation $\precsim$ on a set $M \subseteq \mathbb{R}^n$ is of class C^k on M iff S is a C^k-hypersurface in $\mathbb{R}^n$.

(iv) Convexity. A preference relation $\precsim$ on a convex set M is

(1) weakly convex iff $x \succsim y \Rightarrow \theta x + (1-\theta)y \succsim y \ \forall\ \theta \in [0,1]$;

(2) convex iff $x \succ y \Rightarrow \theta x + (1-\theta)y \succ y \ \forall\ \theta \in (0,1]$;

(3) strongly convex iff $x \sim y$, $x \neq y$, $\Rightarrow \theta x + (1-\theta)y \succ y$ $\forall\ \theta \in (0,1)$.

(v) Completeness. A preference relation $\precsim$ on M is complete iff for any $x,y \in M$, either $x \precsim y$ or $y \precsim x$ or both is the case. □

It is helpful to note that the statements (iv)(1), (2) and (3) are equivalent to: For every $y \in M$, (1) $\bar{S}^+(y)$ is a convex set, (2) $S^+(y)$ is a convex set, (3) $S^+(y)$ is a strictly convex set. Successively, (2) and (3) disallow thick indifference sets and indifference sets with corners. Furthermore, it should be kept in mind that some of the above properties, such as those concerning convexity, may require additional assumptions, such as completeness, in order to make sense in general. Some further discussion of these aspects, including their application in the theory of economic equilibria may be found in Debreu [5].

Together with preference relations on sets it is natural to consider elements which are "most preferred" with respect to these preference relations. In particular, such elements are defined for ordered sets.

Definition 3.5. Optimal elements. Let $(M, \lesssim)$ be a partially ordered [partially preordered] set. Then,

(i) an element $x^0 \in M$ is a minimal element of M iff one has: for every $x \in M$ such that x is comparable to x^0, $x \lesssim x^0 \Rightarrow x = x^0$ $[x \sim x^0]$; an element $x^1 \in M$ is a maximal element iff one has: for every $x \in M$ such that x is comparable to x^1, $x^1 \lesssim x \Rightarrow x^1 = x$ $[x^1 \sim x]$;

(ii) an element $x^0 \in M$ is a least element of M iff $x^0 \lesssim x$ for every $x \in M$, and $x^1 \in M$ is a greatest element iff $x \lesssim x^1$ for every $x \in M$.

Note that minimal (maximal) elements need only be comparable to a subset of M, whereas least (greatest) elements must be comparable to every $x \in M$. For a given $(M, \lesssim)$ none of these elements need exist;

however, when they do, they have more or less obvious properties, some of which are listed next.

M is an ordered set. If the ordering is a partial ordering, then:

(i) a least element (if it exists) is the unique minimal element; however, M may have a unique minimal element, but no least element;

(ii) a least element (if it exists) is unique by antisymmetry;

(iii) it is generally possible to have minimal elements, which are not least elements.

Finally, if $\lesssim$ is a complete (linear) ordering, then the distinction disappears, that is, a minimal element is also a least element.

M is a preordered set. If the preordering is a partial preordering, then:

(i) a least element or a minimal element (if it exists) need not be unique;

(ii) a least element (if it exists) is also a minimal element.

Again, there is no distinction when $\lesssim$ is a complete preordering.

Clearly, these remarks have obvious analogues for maximal (greatest) elements.

The basic assumption for preference optimality is that the criteria space $\mathbb{R}^N$, or a subset thereof, has been preordered by a preference relation $\lesssim$ which is a complete preordering, and which is induced on the set Y or $K(t_1)$. If the basic desire is to minimize the criteria, then the intent is to obtain an element $y^* \in Y$ or $y^* \in K(t_1)$, which is a least element with respect to the preordering $\lesssim$.

A concept which greatly facilitates any calculations with preferences is that of a utility function or order-preserving function; that is, a real-valued function whose numerical values with the usual ordering on $\mathbb{R}$, faithfully reflect the preferences introduced on the set M. In particular, if $\phi(\cdot): M \to \mathbb{R}$ is such a function, then one would like to at least have it satisfy "$x \prec y \Rightarrow \phi(x) < \phi(y)$" for $x,y \in M$. A basic problem in preference and utility theory, however, concerns the conditions on $\precsim$ subject to which there exists such a function $\phi(\cdot)$. That such a function need not exist at all, even when $\precsim$ is a complete preordering, is evidenced by the example of a lexicographic ordering on $\mathbb{R}^2$, which cannot be represented by a utility function. The first appearance of this example in preference literature is ascribed to Debreu [12], although it had been treated earlier in the mathematical literature. An extensive, and well-written, survey of lexicographic orderings is given by Fishburn [13]. For the present treatment it suffices to define a utility function in the following, commonly used manner.

Definition 3.6. Utility function. Let $\precsim$ be a preference relation on a set M. A real-valued function $\phi(\cdot): M \to \mathbb{R}$ is a utility function for $\precsim$ on M iff for every $x,y \in M$:

$$x \precsim y \Leftrightarrow \phi(x) \leq \phi(y),$$

$$x \sim y \Leftrightarrow \phi(x) = \phi(y),$$

$$x \prec y \Leftrightarrow \phi(x) < \phi(y). \quad \square$$

To eliminate any possible confusion in some of the later discussion, the definition of a strictly increasing function is given here.

Definition 3.7. Strictly increasing function. Let $M \subset \mathbb{R}$. A function

$F(\cdot): M \to \mathbb{R}$ is strictly increasing on M iff $x, y \in M$ and $x < y \Rightarrow F(x) < F(y)$. □

Note now that if $F(\cdot)$ is a strictly increasing function on $\mathbb{R}$ and $\phi(\cdot)$ is a utility function for $\precsim$ on a set M, then $\psi(\cdot) = F \circ \phi(\cdot)$ is another utility function for $\precsim$ on M, so that there is nothing unique about a utility function. Once the existence of a utility function has been established it becomes of interest, of course, to discover further conditions on $\precsim$, which allow one to endow $\phi(\cdot)$ with some additional properties such as continuity and differentiability. A usual procedure in consumer theory is to restrict all sets under consideration (such as consumptions sets, etc.) to be subsets of the open positive cone of $\mathbb{R}^N$ defined by

$$P = \{x \in \mathbb{R}^N : x >> 0\}.$$

The set P will recur frequently in this article.

To show the existence of a preference optimal control or decision the continuity of the utility function will be needed. A number of theorems concerning this topic have appeared in the literature with some of the major results being due to Eilenberg [14], Rader [15] and Debreu [16]. It suffices to make use of Eilenberg's theorem. A more refined theorem may be given by introducing additional concepts of utility theory (e.g., see Fishburn [8]). The theorem here is stated for $P \subset \mathbb{R}^N$.

Theorem 3.1. [Eilenberg]. Let $\precsim$ be continuous, and a complete preordering of P. Then there exists a utility function $\phi(\cdot)$ which is continuous on P.† □

---

† Obviously, this is the theorem which gave impetus to the definition of the continuity of a preference relation.

It has already been mentioned that the differentiability of the preference relation is needed in the necessary conditions for preference optimality. It is needed to assure the existence of a differentiable utility function, an assurance which follows from the next theorem, due to Debreu [9].

Theorem 3.2. [Debreu]. The following statements are equivalent; that is, any one of them implies the other two. There exists on P,

(i) a preference relation $\precsim$ which is a complete preordering, weakly monotone, continuous and of class C^2 on P;

(ii) a utility function $\phi(\cdot)$ of class C^2 whose derivative satisfies $\nabla\phi(y) > 0$ for every $y \in P$;

(iii) a C^1-vector field $v(\cdot): P \to P \cap S^N = \{y \in P: \|y\| = 1\}$, with $v(y) > 0$ for every $y \in P$, and such that $v(\cdot)$ satisfies the Frobenius condition

$$\sum_{\substack{i,j,k \\ \text{(cyclic sum)}}} v_i(y)\left(\frac{\partial v_j}{\partial y_k}(y) - \frac{\partial v_k}{\partial y_j}(y)\right) = 0$$

for $i,j,k, = 1,\ldots,N$, $\forall\ y \in P$. □

All of the sufficiency theorems depend on at least the assured pseudo-convexity of a utility function $\phi(\cdot)$ for $\precsim$ on P. This assurance may be available a priori for an explicitly known utility function, it may be implied by conditions imposed on the preference relation, or an arbitrary utility function $\phi(\cdot)$ may be given, for which there exists a convexifying transformation. More specifically, the next theorem provides an answer to the question: Given $\phi(\cdot)$, whose level sets are the indifference surfaces of the preference relation on P, under what conditions on

$\phi(\cdot)$ does there exist a strictly increasing, twice differentiable function $F(\cdot)$ such that $\psi(\cdot) = F\circ\phi(\cdot)$ is a convex utility function for $\precsim$ on P? Without differentiability assumptions this problem was first treated by de Finetti [17] and subsequently by Fenchel [18], who also derived further results based on the inclusion of differentiability assumptions [19]. The results obtained in the last reference are summarized here in the form of a theorem. Whenever possible the notation used by Fenchel has been retained:

For $y \in P$ use

$$\phi_i(y) = \frac{\partial\phi}{\partial y_i}(y) \quad \text{and} \quad \phi_{ij}(y) = \frac{\partial^2\phi}{\partial y_i \partial y_j}(y),$$

to define on P the quadratic form

$$Q(\xi,\xi) = \sum_{i,j=1}^{N} \phi_{ij}(y)\xi_i\xi_j + \sigma(y)\Big(\sum_{i=1}^{N} \phi_i(y)\xi_i\Big)^2 \tag{3.1}$$

where $\sigma(y)$ is a suitable Lagrange multiplier, which one may relate to $F(\cdot)$ by

$$\sigma(y) = \frac{F''(\phi(y))}{F'(\phi(y))}, \tag{3.2}$$

for every $y \in M$. The primes denote differentiation with respect to the argument of $F(\cdot)$. For every $y \in M$ define

$$k^2(y) = \phi_1^2(y) + \phi_2^2(y) + \dots + \phi_N^2(y) \tag{3.3}$$

along with the two characteristic polynomials

$$\Gamma(\alpha) = |\phi_{ij}(y) - \alpha\delta_{ij}| = D_N - D_{N-1}\alpha + \dots + (-1)^N D_0\alpha^N \tag{3.4}$$

$$\Gamma^*(\alpha) = -\frac{1}{k^2(y)}\begin{vmatrix} \phi_{ij}(y) - \alpha\delta_{ij} & \phi_i(y) \\ \phi_j(y) & 0 \end{vmatrix}$$

$$= D^*_{N-1} - D^*_{N-2}\alpha + \ldots + (-1)^{N-1}D^*_0\alpha^{N-1}. \qquad (3.5)$$

The D_ν are the elementary symmetric functions of the characteristic polynomials. With roots $\alpha_1,\ldots,\alpha_N$ of $\Gamma(\alpha)$, and with $D_0 = 1$, these are given by

$$D_1 = \sum_{i=1}^{N}\alpha_i,\quad D_2 = \sum_{\substack{i,j=1\\ i<j}}^{N}\alpha_i\alpha_j,\quad D_3 = \sum_{\substack{i,j,k=1\\ i<j<k}}^{N}\alpha_i\alpha_j\alpha_k,\ldots,\quad D_N = \alpha_1\alpha_2\ldots\alpha_N. \qquad (3.6)$$

With $D^*_0 = 1$, the D^*_ν are similarly defined.

Furthermore, let

$\varepsilon = \inf\{\phi(y): y \in P\}$ and $\beta = \sup\{\phi(y): y \in P\}$

with $\pm\infty$ permitted. The notation (=) in

$$\varepsilon \underset{(=)}{<} t < \beta$$

means that equality holds iff $\phi(\cdot)$ has a minimum on P.

Theorem 3.3. [Fenchel]. Assume that $\phi(\cdot)$ is a twice differentiable function on P. The following conditions are necessary and sufficient for the existence of a twice differentiable strictly increasing function $F(\cdot): \mathbb{R} \to \mathbb{R}$ such that $\psi(\cdot) = F\circ\phi(\cdot)$ is a convex utility function on P:

(i) The function $\phi(\cdot)$ either has no stationary values or it has only an absolute minimum on P.

(ii) The quadratic form $Q(\xi,\xi)$ is positive semidefinite for every $y \in P$.

(iii) If the rank of the matrix of the quadratic form $Q(\xi,\xi)$ is $r(y)-1$ at the point $y \in P$, then the rank of the matrix $[\phi_{ij}(y)]$ is at most $r(y)$.

(iv) For every fixed t with $\varepsilon < t < \beta$

$$G(t) = \sup\{-\frac{D_r}{k^2(y)D^*_{r-1}}: \phi(y) = t,\ y \in P\} < \infty.$$

(v) There exists a function $H(\cdot): [\varepsilon,\beta) \to \mathbb{R}$ such that $H(t) > 0$ on (ε,β), $H(\cdot)$ is differentiable for $\varepsilon < t < \beta$, and such that $G(t) \overset{(=)}{<} \frac{H'(t)}{H(t)}$ (here $H'(t) = \frac{dH}{dt}(t)$). □

The results of the following sections are based on a few general assumptions; for clarity these are given next in a separate section.

4. ASSUMPTIONS

The possible restrictions on the theory, due to these assumptions, are discussed at the end of this section.

Assumption 1. All of the theorems concerning preference optimality are stated with minimization as the basic objective.

Assumption 2. All of the control-theoretic arguments are based on a fixed interval $[t_0,t_1]$ of the independent variable.

Assumption 3. The sets Y and $K(t_1)$ are subsets of P.

Assumption 4. For every $y \in P$ the set $S(y)$ is a hypersurface in $\mathbb{R}^N$, and a normal vector $n(y)$ to $S(y)$ at y is known for every $y \in P$.

The first assumption imposes no restrictions; it is simply a statement to the effect that each criterion function, when considered by itself, is to be minimized. The second assumption imposes no restrictions either, since a control problem with open value t may be transformed to one with fixed interval $[t_0,t_1]$ by means of a simple linear fractional transformation (see Long [20] and Leitmann [21]). The third assumption will not affect any of the work to be done here as long as Y and $K(t_1)$ are bounded below in the coordinatewise ordering, since a translation of the sets will not effect the minimization process. Next, if S(y) is known and S(y) is, say, a C^1-hypersurface in $\mathbb{R}^N$, then a normal vector to S(y) at a given x, $x \in S(y)$ may be constructed.

5. NECESSARY CONDITIONS

It is instructive to introduce the basic concepts by means of a simple example in $\mathbb{R}^2$. It will serve to illustrate the method to be used in the proof of the general maximum principle.

The discussion of this example is intended to give a feeling for the concepts and for the mathematically equivalent notions. The example is purposely kept simple so as not to obscure the relevant points by unnecessary manipulations. The structure of the example makes it desirable to deal with what is basically a maximization problem.

A manufacturer has a production process

$$\dot{x}_1 = u_1, \tag{5.1}$$

which he controls by means of a control $u(\cdot) = (u_1(\cdot),u_2(\cdot))$ with resulting criterion values given by

$$g_1(u(\cdot)) = \int_0^1 x_1(t)\cos u_2(t)dt \quad \text{and} \quad g_2(u(\cdot)) = \int_0^1 x_1(t)\sin u_2(t)dt \tag{5.2}$$

of pairs of leather shoes and leather coats respectively. The process initiates with $x_1(0) = 0$ and the available controls are constrained by $u \in U = \{u \in \mathbb{R}^2 : 2 \leq u_1 \leq 4,\ 0 \leq u_2 \leq \frac{\pi}{2}\}$. Note that $g_1(u(\cdot))$ and $g_2(u(\cdot))$ are always greater than or equal to zero so that $\bar{P}$, the closed positive cone of $\mathbb{R}^2$, suffices as criteria space. Here, $x_1(\cdot)$ may be thought of as machine-use time and the $u_i(\cdot)$ as raw material supplies. Suppose it turns out that for $a \geq 1$ the manufacturer's profit margin remains the same as long as the amounts y_1 and y_2 of pairs of shoes and coats belong to a set

$$S_a = \{y \in \bar{P} : y_2 = \frac{a}{y_1 + 1} - 1\}. \tag{5.3}$$

In addition, however, the parameter a effectively represents the extent to which the process may be used to improve the company's public image; that is, the greater the value a, the more the benefit to the company's image.

Note now that $a_1 \neq a_2 \Rightarrow S_{a_1} \cap S_{a_2} = \emptyset$, and that $\bar{P} = \cup\{S_a : a \geq 1\}$; the collection $\mathcal{D} = \{S_a : a \geq 1\}$ is a decomposition of $\bar{P}$. As such it defines an equivalence relation $\sim$ on $\bar{P}$; that is, if S_a is the equivalence class of $z \in \bar{P}$, then $z^1 \sim z$ iff $z^1 \in S_a$. In still different terms, the set of equivalence classes, $\mathcal{D}$, is sometimes called the quotient $\bar{P}/\sim$ of $\bar{P}$ with respect to the equivalence relation. More

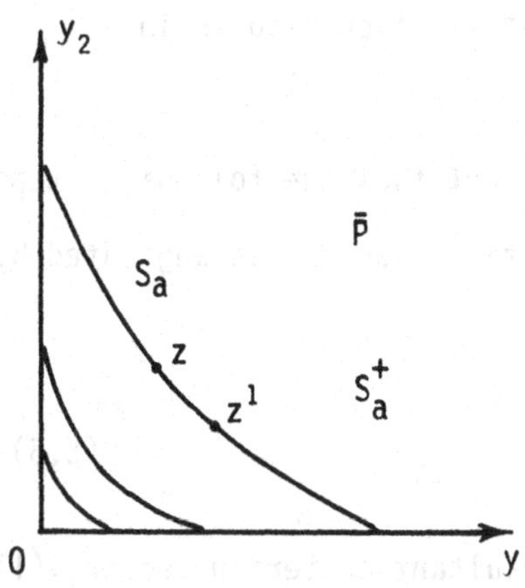

Fig. 2. Indifference sets.

detailed discussions of these equivalent concepts may be found in any basic text on topology, e.g., Willard [22]. Some of these equivalence classes or indifference sets are shown in Figure 2.

The notion of preference follows from the fact that among any two members of the collection $\mathcal{D}$, say S_{a_1} and S_{a_2} with $a_2 \geqslant a_1$, the manufacturer prefers the one with the larger value, a_2; that is, some "points" in the quotient are preferred to others. More precisely, for $a \geqslant 1$ let

$$S_a^+ = \{y \in \bar{P}: y_2 > \frac{a}{y_1 + 1} - 1\} \text{ and } S_a^- = \{y \in \bar{P}: y_2 < \frac{a}{y_1 + 1} - 1\}. \tag{5.4}$$

Then with $z^0 \in S_a$, $z \succ z^0$ (z is preferred to z^0) iff $z \in S_a^+$, and $z \prec z^0$ iff $z \in S_a^-$. Furthermore, $z \precsim z^0$ (z is not preferred to z^0) iff $z \in S_a \cup S_a^-$, and $z \succsim z^0$ iff $z \in S_a \cup S_a^+$. The equivalence relation $\sim$ together with the asymmetric relation $\prec$ constitute a preference relation which is a complete preordering of $\bar{P}$. As before, the preference statements have been made with respect to someone whose basic objective is maximization.

The manufacturer thus seeks a control $u^*(\cdot)$ which results in a greatest element for $\precsim$ on $K(1) \subset \bar{P}$.

In the next section it will become apparent that the following steps are quite generally applicable. First the state space $\mathbb{R}^n$ is augmented by introducing a criterion response

$$\begin{aligned} \dot{y}_1 &= x_1 \cos u_2 \\ \dot{y}_2 &= x_1 \sin u_2 \end{aligned} \tag{5.5}$$

with $y = (y_1, y_2) \in \mathbb{R}^2$, $y(0) = 0$, and with resultant criterion vector $y(1)$. The attainable criteria set (see Figure 3) is

$$K(1) = \{y \in \mathbb{R}^2 : y_2 \geq 1 - y_1,\ y_1^2 + y_2^2 \leq 4,\ y \geq 0\}. \tag{5.6}$$

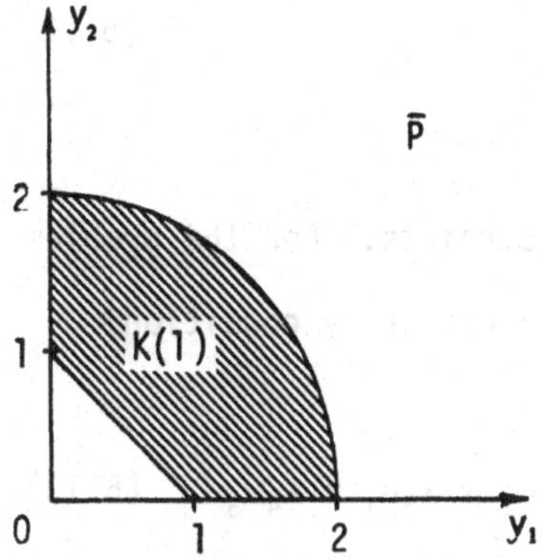

Fig. 3. Attainable criteria set.

The problem in this form is: "Optimize" y(1) subject to

$$\begin{aligned} \dot{y}_1 &= x_1 \cos u_2 \\ \dot{y}_2 &= x_1 \sin u_2 \\ \dot{x}_1 &= u_1 \end{aligned} \tag{5.7}$$

For the above choice of preference relation an obvious choice of a utility function is given by

$$\phi(y) = (y_1 + 1)(y_2 + 1) \tag{5.8}$$

so that one may "optimize" y(1) by maximizing $\phi(y)$ subject to equations (5.7); from $\phi(y) \leq \phi(y^*)$ then follows $y \lesssim y^*$, and the control $u^*(\cdot) \in \mathcal{F}$, which yields $y^*(1) = y^* \in K(t_1)$, is a preference optimal control.

This problem is best dealt with as in Leitmann [21] by further augmenting the system with

$$\dot{x}_2 = u_3,\ x_2(0) = 0, \tag{5.9}$$

subject to $|u_3| < \infty$ with corresponding criterion

$$x_2(1) = \int_0^1 u_3(t)dt, \tag{5.10}$$

and terminal manifold

$$x_2 - \phi(y) = 0. \tag{5.11}$$

With $z = (z_1, z_2, z_3, z_4) = (y_1, y_2, x_1, x_2)$ and with $\tilde{u} = (u_1, u_2, u_3)$ one has:

Maximize $\int_0^1 u_3(t)dt$ subject to

$$\begin{aligned}\dot{z}_1 &= z_3\cos u_2\\ \dot{z}_2 &= z_3\sin u_2\\ \dot{z}_3 &= u_1\\ \dot{z}_4 &= u_3\end{aligned} \tag{5.12}$$

with obvious modifications of the relevant constraints. For the problem in this guise a standard maximum principle is a necessary condition. With $\lambda = (\lambda_0,\lambda_1,\lambda_2,\lambda_3,\lambda_4)$ the H-function is

$$H(\lambda,z,\tilde{u}) = \lambda_0 u_3 + \lambda_1 z_3 \cos u_2 + \lambda_2 z_3 \sin u_2 + \lambda_3 u_1 + \lambda_4 u_3 \tag{5.13}$$

and the adjoint equations are

$$\begin{aligned}\dot{\lambda}_i &= 0 \text{ for } i = 0,1,2,4\\ \dot{\lambda}_3 &= -\lambda_1\cos u_2 - \lambda_2\sin u_2,\end{aligned} \tag{5.14}$$

so that $\lambda_i(t) = d_i$ for $i = 0,1,2,4$, where the d_i's are constants. Now let $\tilde{u}^*(\cdot)$ be an optimal control for the problem above, and let $z^*(\cdot)$ be a corresponding solution of the state equations (5.12). Then there necessarily exists a nonzero response $\lambda^*(\cdot)$ of equations (5.14) evaluated at $(z^*(t),u^*(t))$, with $\lambda_0^*(t) = d_0^* = \lambda_0^* = \text{const.} \geq 0$. Since u_3 is unconstrained

$$\frac{\partial H}{\partial u_3}(\lambda^*(t),z^*(t),\tilde{u}^*(t)) = 0 \tag{5.15}$$

is necessary, and $\lambda_4^*(t) = d_4^* = -\lambda_0^*$ follows. The initial transversality conditions are satisfied identically; with η in the tangent space to the terminal manifold, the terminal transversality conditions are given by

$$\sum_{i=1}^{4} \lambda_i^*(1)\,\eta_i = 0, \tag{5.16}$$

subject to

$$\eta_4 - \sum_{i=1}^{2} \frac{\partial\phi}{\partial y_i}(y^*(1))\,\eta_i = 0. \tag{5.17}$$

Consequently,

$$\eta_1(\lambda_1^*(1) - \lambda_0^*\frac{\partial\phi}{\partial y_1}(y^*(1)) + \eta_2(\lambda_2^*(1) - \lambda_0^*\frac{\partial\phi}{\partial y_2}(y^*(1)) + \lambda_3(1)\eta_3 = 0, \tag{5.18}$$

which implies $\lambda_3^*(1) = 0$ and

$$\lambda_1^*(t) = d_1^* = \lambda_1^*(1) = \lambda_0^*\frac{\partial\phi}{\partial y_1}(y^*(1)) = \lambda_0^* c_1,$$
$$\lambda_2^*(t) = d_2^* = \lambda_2^*(1) = \lambda_0^*\frac{\partial\phi}{\partial y_2}(y^*(1)) = \lambda_0^* c_2. \tag{5.19}$$

The evaluation of the c_i in terms of the $\frac{\partial\phi}{\partial y_i}(y^*(1))$ is postponed; for the moment only the fact $\nabla\phi(y) > 0 \ \forall\ y \in P$ is used to conclude that $c > 0$. With these results the H-function may be written as

$$H(\lambda^*(t), z^*(t), \tilde{u}) = \lambda_0^* |z_3^*(t)| \sqrt{c_1^2 + c_1^2} \sin\left(\tan^{-1}\frac{c_1}{c_2} + u_2\right) + \lambda_3^*(t) u_1. \tag{5.20}$$

The $\sup_{\tilde{u}\in\tilde{U}} H(\lambda^*(t), z^*(t), \tilde{u})$ for u_2 is attained when

$$\tan^{-1}\frac{c_1}{c_2} + u_2^* = \frac{\pi}{2} \tag{5.21}$$

or $u_2^*(t) = \tan^{-1}\frac{c_2}{c_1}$. Substitution of this result in the second of the adjoint equations (5.14) yields $\lambda_3(t) \geq 0 \ \forall\ t \in [0,1]$, and a control $\tilde{u}^*(\cdot)$ satisfying the necessary conditions is defined by

$$\tilde{u}^*(t) = (4, \tan^{-1}\frac{c_2}{c_1}, u_3(t)) \tag{5.22}$$

where $u_3(\cdot)$ may be any arbitrary, bounded and Lebesgue measurable control function. It follows that the control $u^*(\cdot)$ defined by

$$u^*(t) = (4, \tan^{-1}\frac{c_2}{c_1}) \tag{5.23}$$

with $c > 0$ is a candidate for preference optimal control.

Note now that the following two part problem is equivalent to the development just given. As a first step, let the H-function be given by

$$\hat{H}(\hat{\lambda}, \hat{x}, u) = \lambda_0(c_1 x_1 \cos u_2 + c_2 x_1 \sin u_2) + \lambda_1 u_1 \tag{5.24}$$

where $\hat{\lambda} = (\lambda_0,\lambda_1)$, $\hat{x} = (y_1,y_2,x_1)$ and where the adjoint equations are

$$\dot{\lambda}_0 = 0, \qquad \dot{\lambda}_1 = - c_1\cos u_2 - c_2\sin u_2. \tag{5.25}$$

In terms of these one then has a modified maximum principle, where the usual single criterion integrand has been replaced by a scalar combination of criterion integrands with weighting constants $(c_1,c_2) > 0$. Up to this point only the fact $\nabla\phi(y) > 0 \ \forall\ y \in K(t_1)$ has been used.

An additional condition, given later, consists of the imposition of a compatibility condition of the form

$$c = n(y^*(1)), \tag{5.26}$$

where $n(y^*(1))$ is normal to $S(y^*(1))$ at $y^*(1)$.

The central ideas of this section have been presented in the preceding example. The mathematical concepts will now be made precise. The results of this section, although stated for multiple criteria problems, are equally valid for cooperative games; in particular, they are also valid, when the number of control components $r < N$, the number of criteria. Before proceeding to a maximum principle the definition of preference optimality is given together with a brief discussion of existence.

<u>Definition 5.1</u>. Preference optimal control. Let $K(t_1) \subset P$ and let $\precsim$ be a complete preordering of P. A control $u^*(\cdot) \in \mathcal{F}$ is a preference optimal control iff it results in a corresponding solution $x^*(\cdot)$ such that the pair $\{u^*(\cdot),x^*(\cdot)\}$ generates a criterion response $y^*(\cdot)$ for which $y^*(t_1) = y^*$ is a least element of $K(t_1)$ with respect to $\precsim$. □

To prove existence, only the continuity of the utility function is needed. Thus Theorem 3.1 is used.

Theorem 5.1. Let $K(t_1)$ (compact) $\subset P$ and let $\precsim$ be continuous and a complete preordering of P. Then there exists a preference optimal control $u^*(\cdot)$.

Proof: In view of Theorem 3.1 there exists a continuous utility function $\phi(\cdot)$ representing $\precsim$ on P. Since $K(t_1)$ is compact there exists a $y^* \in K(t_1)$ such that $\phi(y^*) \leq \phi(y)$ for every $y \in K(t_1)$. This implies $y^* \precsim y \;\forall\; y \in K(t_1)$, that is, y^* is a least element for $\precsim$ on $K(t_1)$. With $y^* \in K(t_1)$ there exists an admissible control $u^*(\cdot)$ which generates a $y^*(\cdot)$ with terminal point $y^* = y^*(t_1)$. The control $u^*(\cdot)$ is preference optimal. ■

In general, existence may be treated as in a terminal criterion optimal control problem or as in a standard programming problem.

To the extent needed, the statement of the following maximum principle is based on Lee and Markus [4].

Theorem 5.2. Consider a control process

$$\dot{x} = f(x,u) \tag{5.27}$$

with $f(\cdot)$ and $\frac{\partial f}{\partial x}(\cdot)$ continuous on $\mathbb{R}^{n+r}$ and let $u(\cdot) \in \mathcal{F}$. For each such $u(\cdot)$ with response $x(\cdot)$ let the criteria be

$$g_i(u(\cdot)) = \int_{t_0}^{t_1} f_{0i}(x(t),u(t))dt, \tag{5.28}$$

$i = 1, \ldots, N$, with $f_{0i}(\cdot)$ and $\frac{\partial f_{0i}}{\partial x}(\cdot)$ continuous on $\mathbb{R}^{n+r}$. Let the preference relation $\precsim$ on P be a complete preordering, continuous, weakly monotone, and of class C^2.

If $u^*(\cdot)$ is preference optimal in $\mathcal{F}$ with response $\hat{x}^*(\cdot) = (x^*(\cdot),y^*(\cdot))$, then there exists a vector $c \in \mathbb{R}^N$, $c > 0$ such that

$$\hat{H}(\hat{\lambda},\hat{x},u) = \lambda_0 \sum_{i=1}^{N} c_i f_{0i}(x,u) + \sum_{k=1}^{n} \lambda_k f_k(x,u) \tag{5.29}$$

with $\hat{\lambda} = (\lambda_0,\lambda_1,\ldots,\lambda_n) = (\lambda_0,\lambda)$, and the adjoint equations

$$\dot{\lambda}_r = - \lambda_0 \sum_{i=1}^{N} c_i \frac{\partial f_{0i}}{\partial x_r}(x,u) - \sum_{k=1}^{n} \lambda_k \frac{\partial f_k}{\partial x_r}(x,u), \tag{5.30}$$

$r = 0, 1, \ldots, n$, satisfy the following conditions:

There exists a nontrivial response $\hat{\lambda}^*(\cdot): [t_0,t_1] \to C(\text{open}) \subset \mathbb{R}^{1+n}$ of equation (5.30) evaluated at $(x^*(\cdot),u^*(\cdot))$, with $\lambda_0^*(t) = \lambda_0^* = \text{const.} \leq 0$ everywhere on $[t_0,t_1]$ and with

$$\sup_{u \in U} \hat{H}(\hat{\lambda}^*(t),\hat{x}^*(t),u) = \hat{H}(\hat{\lambda}^*(t),\hat{x}^*(t),u^*(t)) = 0 \tag{5.31}$$

almost everywhere on $[t_0,t_1]$. Also, if θ^0 and θ^1 are manifolds with tangent spaces T_0 and T_1 at $x^*(t_0)$ and $x^*(t_1)$, then

$$\begin{aligned} &\lambda^*(t_0) \text{ is orthogonal to } T_0, \\ &\lambda^*(t_1) \text{ is orthogonal to } T_1. \end{aligned} \tag{5.32}$$

Proof: Augment the state space by introducing

$$\dot{y} = f^0(x,u) \tag{5.33}$$

with $y(t_0) = 0$. By Assumption 3 the attainable criteria set $K(t_1) \subset P$, and by the hypotheses above $\lesssim$ is a weakly monotone, continuous preference relation of class C^2 which completely preorders P. According to Theorem 3.2 there thus exists a C^2-utility function $\phi(\cdot)$ representing $\lesssim$ on P. Hence, the problem becomes:

$$\begin{aligned} &\text{Minimize } \phi(y(t_1)) \text{ subject to} \\ &\dot{y} = f^0(x,u),\ y(t_0) = 0 \\ &\dot{x} = f(x,u),\ x(t_0) \in \theta^0 \text{ and } x(t_1) \in \theta^1. \end{aligned} \tag{5.34}$$

Again, introduce

$$x_{n+1}(t_1) = \int_{t_0}^{t_1} u_{r+1}(t)dt = \phi(y(t_1)) \tag{5.35}$$

with

$$\dot{x}_{n+1} = u_{r+1} \text{ and } x_{n+1}(t_0) = 0, \tag{5.36}$$

where $u_{r+1}(\cdot)$ is any bounded Lebesgue measurable function, a condition which is denoted by $|u_{r+1}| < \infty$.† In this guise the problem is:

Minimize

$$\int_{t_0}^{t_1} u_{r+1}(t)dt \tag{5.37}$$

subject to

$$\begin{aligned} \dot{y} &= f^0(x,u),\ y(t_0) = 0, \\ \dot{x} &= f(x,u),\ \ x(t_0) \in \theta^0 \text{ and } x(t_1) \in \theta^1, \\ \dot{x}_{n+1} &= u_{r+1},\ x_{n+1}(t_0) = 0 \text{ and } x_{n+1}(t_1) - \phi(y(t_1)) = 0. \end{aligned} \tag{5.38}$$

For this problem the maximum principle in its usual form is a necessary condition. Let $\tilde{x} = (y,x,x_{n+1}) \in \mathbb{R}^{N+n+1}$, $\tilde{u} = (u,u_{r+1}) \in \tilde{U} \subset \mathbb{R}^{r+1}$, $\tilde{\lambda} = (\psi,\hat{\lambda},\lambda_{n+1}) \in \mathbb{R}^{N+n+2}$ and define

$$\tilde{H}(\tilde{\lambda},\tilde{x},\tilde{u}) = \lambda_0 u_{r+1} + \sum_{i=1}^{N} \psi_i f_{0i}(x,u) + \sum_{k=1}^{n} \lambda_k f_k(x,u) + \lambda_{n+1} u_{r+1} \tag{5.39}$$

with adjoint equations $\dot{\lambda}_0 = 0$ and

$$\dot{\lambda}_r = - \sum_{i=1}^{N} \psi_i \frac{\partial f_{0i}}{\partial x_r}(x,u) - \sum_{k=1}^{n} \lambda_k \frac{\partial f_k}{\partial x_r}(x,u) \tag{5.40}$$

† Problems where the control may be any bounded measurable function are sometimes called free or unconstrained.

for r = 1, ..., n, and

$$\dot{\lambda}_{n+1} = 0,$$

$$\dot{\psi}_s = 0,$$

for s = 1, ..., N. If $\tilde{u}^*(\cdot)$ is an optimal control for this last problem and $\tilde{x}^*(\cdot)$ the corresponding response, then there exists a nontrivial solution of the adjoint equations (5.40) and (5.41) such that

$$\sup_{\tilde{u}\in\tilde{U}} \tilde{H}(\tilde{\lambda}^*(t),\tilde{x}^*(t),\tilde{u}) = \tilde{H}(\tilde{\lambda}^*(t),\tilde{x}^*(t),\tilde{u}^*(t)) = 0^\dagger \quad (5.42)$$

and $\lambda_0^*(t) = \lambda_0^* = \text{const.} \leq 0$. Since $u_{r+1}(\cdot)$ is unconstrained

$$\frac{\partial\tilde{H}}{\partial u_{r+1}}(\tilde{\lambda}^*(t),\tilde{x}^*(t),\tilde{u}^*(t)) = 0 \quad 5.43)$$

is necessary, with the result

$$\lambda_{n+1}^*(t) = -\lambda_0^* = \text{const.} \geq 0. \quad (5.44)$$

At $t = t_0$ the transversality conditions are

$$\lambda^*(t_0) \perp T_0 \quad (5.45)$$

since the remaining ones are satisfied identically. The terminal transversality condiuons are given by

$$\lambda^*(t_1) \perp T_1 \quad (5.46)$$

and

$$\lambda_{n+1}^*(t_1)\, \eta_{N+1} + \sum_{i=1}^{N} \psi_i^*(t_1)\eta_i = 0, \quad (5.47)$$

where the η_i's satisfy

† Recall that the formulation involves $x_n = t$.

$$\eta_{N+1} - \sum_{i=1}^{N} \frac{\partial \phi}{\partial y_i}(y^*(t_1))\eta_i = 0. \tag{5.48}$$

With (5.44) and (5.48) equation (5.47) becomes

$$\sum_{i=1}^{N} (- \lambda_0^* \frac{\partial \phi}{\partial y_i}(y^*(t_1)) + \psi_i^*(t_1))\eta_i = 0. \tag{5.49}$$

Since there are no further restrictions on the η_i it follows that

$$\psi_i^*(t_1) = \lambda_0^* \frac{\partial \phi}{\partial y_i}(y^*(t_1)) = \lambda_0^* c_i, \tag{5.50}$$

or, in view of (5.41),

$$\psi_i^*(t) = \lambda_0^* c_i \tag{5.51}$$

for i = 1, ..., N, where the $c_i \geq 0$, not all zero, as a consequence of $\nabla\phi(y) > 0$.

In light of these results one may just as well consider

$$\hat{H}(\hat{\lambda},\hat{x},u) = \lambda_0 \sum_{i=1}^{N} c_i f_{0i}(x,u) + \sum_{k=1}^{n} \lambda_k f_k(x,u) \tag{5.52}$$

along with the adjoint equations $\dot{\lambda}_0 = 0$ and

$$\dot{\lambda}_r = - \lambda_0 \sum_{i=1}^{N} c_i \frac{\partial f_{0i}}{\partial x_r}(x,u) - \sum_{k=1}^{n} \lambda_k \frac{\partial f}{\partial x_r}(x,u), \tag{5.53}$$

r = 1, ..., n, and with transversality conditions

$\lambda(t_0) \perp T_0$ and $\lambda(t_1) \perp T_1$. ■

In the case $\lambda_0 < 0$ one may set $\lambda_0 = -1$ as usual; in the abnormal case $\lambda_0 = 0$ and the H-function becomes independent of the criteria integrands, a situation which is analogous to that in an optimal control problem.

Similar results are now obtained for the programming problem.

Definition 5.2. Preference optimal decision. Let $\lesssim$ be a complete preordering of Y. A decision $x^* \in X$ is preference optimal iff it results in a $y^* = g(x^*)$ such that y^* is a least element for $\lesssim$ on Y. □

The following theorem deals with necessary conditions for a preference optimal decision. In essence, the theorem is a corollary to Theorem 5.2.

Theorem 5.3. Let $h(\cdot)$ and $g(\cdot)$ have continuous first partial derivatives at x^*, and let $f(\cdot)$ be differentiable at x^*. Let $\lesssim$ be a complete preordering, continuous, weakly monotone and of class C^2 on P. If x^* is a preference optimal decision on X, then there exist vectors $c \in \mathbb{R}^N$, $(\lambda_0,\lambda) \in \mathbb{R}^{1+m}$ and $\mu \in \mathbb{R}^k$ such that

$$\begin{aligned}
\lambda_0 c\nabla g(x^*) + \lambda\nabla f(x^*) + \mu\nabla h(x^*) &= 0,\\
f(x^*) &\leqslant 0,\\
\lambda f(x^*) &= 0,\\
h(x^*) &= 0,\\
c &> 0,\\
(\lambda_0,\lambda) &\geqslant 0,\\
(\lambda_0,\lambda,\mu) &\neq 0.
\end{aligned} \tag{5.54}$$

Proof: Let $y = (g_1,\dots,g_N) \in P$, and let $z = (x,y) \in \Omega \times P$. As a consequence of the assumptions concerning the preference relation $\lesssim$, there exists on P C^2-utility function $\phi(\cdot)$ with $\nabla\phi(y) > 0\ \forall\ y \in P$. One may thus formulate the problem in terms of z as

minimize $\tilde{\phi}(z)$ subject to $z \in Z$,

where

$$Z = \{z \in \Omega\times P:\ \tilde{f}(z) \leqslant 0,\ \tilde{h}(z) = 0\} \tag{5.55}$$

with mappings

$$\tilde{f}(\cdot): \Omega\times P \to \mathbb{R}^m \quad \text{and} \quad \tilde{h}(\cdot): \Omega\times P \to \mathbb{R}^{k+N} \tag{5.56}$$

defined by

$$\begin{aligned} &\tilde{f}_j(z) = f_j(x), \quad j = 1,\ldots,m \\ &\tilde{h}_j(z) = \begin{cases} h_j(x), & j = 1,\ldots,k \\ y_j - g_j(x), & j = k+1,\ldots,k+N. \end{cases} \end{aligned} \tag{5.57}$$

This statement now is an ordinary, nonlinear programming problem with minimization as the objective. A necessary condition for a minimizing $z^* = (x^*,y^*) \in Z$ is the Fritz John stationary point condition.[7] For such a z^*, there then exists a

$$\tilde{\lambda} = (\lambda_0,\lambda,\mu,\psi) \in \mathbb{R}^{1+m+k+N}$$

such that

$$\begin{aligned} \lambda_0\nabla_z\tilde{\phi}(z^*) + \lambda\nabla_z\tilde{f}(z^*) + (\mu,\psi)\nabla_z\tilde{h}(z^*) &= 0, \\ \tilde{h}(z^*) = 0, \quad \tilde{f}(z^*) &\leq 0, \\ \lambda\tilde{f}(z^*) &= 0, \\ (\lambda_0,\lambda) &\geq 0, \\ \tilde{\lambda} &\neq 0. \end{aligned} \tag{5.58}$$

In terms of x and y this condition becomes

$$\begin{aligned} \lambda\nabla_x f(x^*) + \mu\nabla_x h(x^*) - \psi\nabla_x g(x^*) &= 0, \\ \lambda_0\nabla_y\phi(y^*) + \psi &= 0. \end{aligned} \tag{5.59}$$

Suppression of the subscript x on ∇_x along with

$$\psi = -\lambda_0\nabla_y\phi(y^*) = -\lambda_0 c, \quad c > 0, \tag{5.60}$$

then yields the equivalent condition

$$\lambda_0 c\nabla g(x^*) + \lambda\nabla f(x^*) + \mu\nabla h(x^*) = 0. \tag{5.61}$$

The conclusion of the theorem follows by noting that if $(\lambda_0,\lambda,\mu) = 0$, then $\lambda_0 = 0$, which would imply $\psi = 0$, a contradiction to $\tilde{\lambda} \neq 0$. ■

As is the case for the general nonlinear programming problem one may assure $\lambda_0 > 0$ by imposing constraint qualifications.

Hitherto, the necessary conditions in part provided assurance that if a control or decision is preference optimal then there exists a vector $c > 0$, $c \in \mathbb{R}^N$ such that a scalar combination of the criteria satisfies a maximum principle or a Fritz John condition. As was noted in the earlier example only $\nabla\phi(y) > 0 \ \forall \ y \in P$ was used to deduce $c > 0$. As of yet no use has been made of the fact that $\nabla\phi(y)$ is normal to $S(y)$ at y; this will impose an additional restriction on c. Without this restriction the previously derived conditions are no different from those for non-domination, or a special case thereof, Pareto optimality (see Yu [23] and Yu and Leitmann [24]. Before proceeding to the compatibility condition which c must satisfy, a closer look at the influence of the vector c is warranted.

Quite generally, let

$$\hat{H}(\hat{\lambda},\hat{x},u) = \lambda_0 \sum_{i=1}^{N} c_i f_{0i}(x,u) + \sum_{k=1}^{n} \lambda_k f_k(x,u), \tag{5.62}$$

where the c_i are arbitrary constants. Consider any solution concept resulting in necessary conditions which involve a maximum principle stated in terms of the just given H-function. A control $u^*(\cdot)$, satisfying necessary conditions, generally will depend on the parameters c_i, and this dependence will be induced on the adjoint variables, the solution to the state equations and the criterion functions. With a slight abuse of notation denote this dependence by

$u^*(c;\cdot)$, $\lambda^*(c;\cdot)$, $x^*(c;\cdot)$ and $y^*(c;\cdot)$, with $c \in \mathbb{R}^N$.

The solution concepts themselves then differ only in the permissible values or range of values of the c_i. For preference optimality with the particular preference relation on which the necessary conditions have been based and for Pareto-optimality it follows from the respective necessary conditions that c belongs to the set

$$C = \{c \in \mathbb{R}^N: c > 0\}.$$

Furthermore, in a strict sense all the quantities depend only on the relative ratios of the c_i, since none of the results are altered if the H-function is divided by a non-zero constant, a fact which will be illustrated shortly. Any further restriction of the values of the c_i due to necessary conditions depends on the optimality concept to be used. For preference optimality this additional selection process is embodied in the following lemma.

Lemma 5.1. Assume that there is on P a preference relation $\lesssim$ which is a complete preordering, continuous, weakly monotone and of class C^2 on P. Let $c^* \in C$ be such that $u^*(c^*;\cdot) \in \mathfrak{F}$ is a preference optimal control with corresponding $x^*(c^*;\cdot)$ and $y^*(c^*;\cdot)$. Let $y^* = y^*(c^*;t_1)$ and assume that $S(y^*)$ is a C^1-hypersurface in P. Then there exists a normal vector $n(y^*)$ to $S(y^*)$ at y^*, which may be chosen to point into $\overline{S}^+(y^*)$ and for any such normal vector one may choose

$$c^* = n(y^*).$$

Proof: Due to the assumptions on $\lesssim$ there exists on P a utility function $\phi(\cdot)$ with $\nabla\phi(y) > 0 \ \forall \ y \in P$. Thus, $\nabla\phi(y^*) > 0$ and any vector $n(y^*) = k\nabla\phi(y^*)$, $k > 0$, is an appropriate normal vector. The rest follows from the definition of c in the preceding theorem and from the fact that the H-function is insensitive to division by a non-zero constant. ■

Naturally there is a similar condition for the programming problem.

The lemma is now illustrated in terms of the example above. As a result of the application of the maximum principle one had

$$u^*(c;t) = (4, \tan^{-1}\frac{c_2}{c_1}) \tag{5.64}$$

together with

$$y_1^*(c;1) = \frac{2c_1}{\sqrt{c_1^2 + c_2^2}} \quad \text{and} \quad y_2^*(c;1) = \frac{2c_2}{\sqrt{c_1^2 + c_2^2}} . \tag{5.65}$$

For any $y \in P$ a normal vector to $S(y)$ at y is given by

$$n(y) = (y_2 + 1,\ y_1 + 1) \tag{5.66}$$

so that one has, in particular,

$$n(y^*(c;1)) = (1 + \frac{2c_2}{\sqrt{c_1^2 + c_2^2}}, 1 + \frac{2c_1}{\sqrt{c_1^2 + c_2^2}}). \tag{5.67}$$

The compatibility condition (Lemma 5.1) then requires

$$c_1 = 1 + \frac{2c_2}{\sqrt{c_1^2 + c_2^2}}, \quad c_2 = 1 + \frac{2c_1}{\sqrt{c_1^2 + c_2^2}}, \tag{5.68}$$

whose simultaneous solution yields $c_1^* = c_2^*$. Corresponding to this condition one has finally

$$y_1^*(c^*;1) = y_2^*(c^*;1) = \sqrt{2} \tag{5.69}$$

resulting from a control

$$u^*(c^*;t) = (4, \frac{\pi}{4}). \tag{5.70}$$

There are some additional consequences due to the restriction of the c_i. For $c \in C$ and $u^*(c;\cdot)$ a subset of $K(t_1)$ is generated, namely

$$K^*(c;t_1) = \{y \in \mathbb{R}^N:\ y = y^*(c;t_1),\ c \in C\}. \tag{5.71}$$

Lemma 5.2. Let $c^* \in C$ be such that $u^*(c^*;\cdot) \in \mathfrak{F}$ is a preference optimal control and let $\phi(\cdot)$ be any utility function for $\lesssim$ on P. Furthermore, let $y^*(c^*;\cdot)$ be the criterion response generated by $u^*(c^*;\cdot)$. Then

$$\phi(y^*(c^*;t_1)) \leq \phi(y^*(c;t_1)) \tag{5.72}$$

for every $c \in C$.

Proof: If $u^*(c^*;\cdot)$ is preference optimal and $y^* = y^*(c^*;t_1)$ is a corresponding criterion response point, then $\phi(y^*) \leq \phi(y)\ \forall\ y \in K(t_1)$ and hence for $y \in K^*(c;t_1)$. ■

Thus it is necessary that $\tilde{\phi}(\cdot)$ considered as a function of c be minimized with respect to c.

Recall now that maximization was the basic objective in the example above so that an equivalent result when applied to the example yields: Maximize $\tilde{\phi}(\cdot)$ subject to $c \in C$, where

$$\tilde{\phi}(c) = \left(1 + \frac{2c_1}{\sqrt{c_1^2 + c_2^2}}\right)\left(1 + \frac{2c_2}{\sqrt{c_1^2 + c_2^2}}\right). \tag{5.73}$$

A necessary condition for a maximum is

$$\frac{\partial\tilde{\phi}}{\partial c_1}(c) = \frac{2c_2}{(c_1^2 + c_2^2)^2}[(c_1^2 + c_2^2)^{1/2} + 2(c_1 + c_2)](c_2 - c_1)\Big|_{c=c^*} = 0, \tag{5.74}$$

with a similar condition for c_2, and $c_1^* = c_2^*$ follows. It is clear from the present example that a normalization condition could have been included. In the present case this is expressed by the fact that $\phi(\cdot)$ depends only on the ratio $\theta = \frac{c_2}{c_1}$. With $0 \leq \theta \leq \infty$ an equivalent formulation then is

$$\frac{d\hat{\phi}}{d\theta}(\theta) = \frac{2}{(1+\theta^2)^2}[\sqrt{1+\theta^2} + 2(1+\theta)](1-\theta)\Big|_{\theta=\theta^*} = 0, \tag{5.75}$$

from which $\theta^* = 1$ follows.

The last lemma may be used in a still different manner. As has already been indicated the controls $u^*(c;\cdot)$, $c \in C$, generate a subset $K^*(c;t_1)$ of $K(t_1)$. The dependence on the parameters c_i may be suppressed in the description of this set; for the present example this results in

$$K^*(c;1) = \{y \in \mathbb{R}^2 : y_1^2 + y_2^2 = 4,\ y > 0\}. \tag{5.76}$$

One then has the programming problem: Maximize $\phi(y)$ subject to $y \in K^*(c;1)$. This equality constraint may be substituted in $\phi(\cdot)$ directly, and a necessary condition is again

$$\frac{d\bar{\phi}}{dy_2}(y_2) = (4-y_2^2)^{-1/2}[4 + (4-y_2^2)^{1/2} - 2y_2^2 - y_2]\Big|_{y_2=y_2^*} = 0 \tag{5.77}$$

from which $y_2^* = \sqrt{2}$ and consequently $c_1^* = c_2^*$ follow.

6. SUFFICIENT CONDITIONS

Briefly, if preference optimal control can be shown to exist, and if the solution to the compatibility conditions (Lemma 5.1), subject to a normalization constraint is unique, then the resultant control $u^*(c^*;\cdot)$ is preference optimal. In the example $K(t_1)$ is compact, so that the existence theorem applies; $c_1^* = c_2^*$ is the unique solution of the compatibility relation, so that one may conclude that

$$u^*(t) = (4, \tfrac{\pi}{4})$$

defines a preference optimal control. Note that in this case no explicit

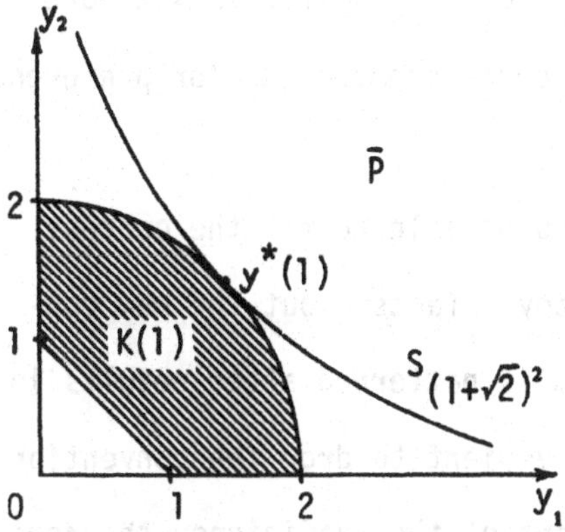

Fig. 4. Preference optimal criterion value and its indifference class $S_{(1+\sqrt{2})^2}$

knowledge of the utility function was needed. Now if u_1 happens to denote leather in lots of a hundred hides, u_2 denotes thread in ten-thousands of yards, machine-use time x_1 is in hundreds of hours, y_1 is in hundreds of pairs of shoes and y_2 denotes the number of leather coats in hundreds, then the answer to the initially posed physical problem is: Corresponding to a resource use of 400 hides and 7850 yards of thread a machine-use time of 400 hours results in 141 coats and 141 pairs of shoes. Naturally it has been assumed for all quantities that they are perfectly divisible. The result is depicted in Figure 4.

This situation is analogous to arguments in optimal control and programming, that is, if an optimal control is known to exist and the control satisfying a maximum principle is unique, then it is the optimal control. Recall that the necessary conditions involved no explicit knowledge of a utility function, only its existence along with suitable properties needed to be assured. Ideally, sufficient conditions should also be derived based only on hypotheses in terms of the preference relation. This possibility is treated at the end of this section, where an indication is given concerning a strengthening of Debreu's Theorem to yield a convex, C^2 utility function. The first two theorems here are based on the explicit knowledge of a utility function, followed by a general theorem which may be used to transform any sufficiency theorem in optimal control or programming into a corresponding one for preference optimality.

In any new concept it is desirable to be able to tie the concept into existing theory and theorems to discover facts about the new concept. The proof of the following theorem is patterned after results in Lee and Markus [4]. Furthermore, it is convenient to drop the convention $x_n = t$ for this theorem. For the statement of the sufficiency theorem, introduce the mappings

$$\begin{aligned} &k^0(\cdot): A \times [t_0,t_1] \to K^0(\text{open}) \subset \mathbb{R}^N, \quad k^0 = (k_{01},\ldots,k_{0N}),\\ &h^0(\cdot): U \times [t_0,t_1] \to H^0(\text{open}) \subset \mathbb{R}^N, \quad h^0 = (h_{01},\ldots,h_{0N}), \qquad (6.1)\\ &h(\cdot): U \times [t_0,t_1] \to H(\text{open}) \subset \mathbb{R}^n, \quad h = (h_1,\ldots,h_n), \end{aligned}$$

and an n×n matrix A(t) to define

$$f^0(x(t),t,u(t)) = k^0(x(t),t) + h^0(u(t),t) \qquad (6.2)$$

and

$$f(x,t,u) = A(t)x + h(u,t). \qquad (6.3)$$

<u>Theorem 6.1</u>. Let $u^*(\cdot) \in \mathcal{F}$ and let $\hat{x}^*(\cdot)$ be the corresponding augmented response. Let $\precsim$ be a preference relation on P which is a complete preordering of P. Assume:

(i) $S(y^*)$ is such that $n(y^*)$ can be chosen to satisfy $n(y^*) > 0$.

(ii) A utility function $\phi(\cdot)$ is known for $\precsim$ on P.

(iii) The gradient $\nabla\phi(y^*) > 0$ and $\phi(\cdot)$ satisfies the hypotheses of Theorem 3.3.

(iv) $k_{0i}(\cdot)$, $\nabla k_{0i}(\cdot)$, $h_{0i}(\cdot)$, $i = 1, \ldots, N$, along with $A(\cdot)$ and $h(\cdot)$ are continuous on their respective domains.

(v) The $k_{0i}(\cdot)$, $i = 1, \ldots, N$ are convex in x for each fixed t in the finite interval $[t_0,t_1]$.

(vi) $\theta^0 = \{x^0\}$, a single point, and $\theta^1 \subseteq \mathbb{R}^n$ is a closed convex set.

(vii) The control $u^*(\cdot)$ satisfies the maximum principle

$$- ch^0(u^*(t),t) + \lambda(t)h(u^*(t),t) = \max_{u \in U}(- ch^0(u,t) + \lambda(t)h(u,t)), \quad (6.4)$$

where $c > 0$ and satisfies the compatibility condition $c = n(y^*)^\dagger$, and where $\lambda(t)$ is any nontrivial solution of

$$\dot{\lambda}(t) = c\nabla k^0(x^*(t),t) - \lambda(t)A(t) \quad (6.5)$$

satisfying the transversality condition:

$\lambda(t_1)$ is an inward normal to θ^1 at the boundary point $x^*(t_1)$.

Then $u^*(\cdot)$ is a preference optimal control.

Proof: Similar to the procedure in Lee and Markus [4] it is first shown that the inequality

$$- c(y(t_1) - y^*) + \lambda(t_1)(x(t_1) - x^*(t_1)) \geqslant 0 \quad (6.6)$$

holds. Consider the derivative

$$\frac{d}{dt}[- cy(t) + \lambda(t)x(t)] = - ck^0(x(t),t) + c\nabla k^0(x^*(t),t)x(t) - ch^0(u(t),t) + \lambda(t)h(u(t),t). \quad (6.7)$$

† Naturally, this condition is to be interpreted in the sense of Lemma 5.1.

The integration of both sides with respect to t results in

$$- cy(t_1) + [\lambda(t_1)x(t_1) - \lambda(t_0)x(t_0)] = \int_{t_0}^{t_1} \{[- ck^0(x(t),t) + c\nabla k^0(x^*(t),t))x(t)] + [- ch^0(u(t),t) + \lambda(t)h(u(t),t)]\}dt, \tag{6.8}$$

since $y(t_0) = 0$. The same expression evaluated at $u^*(\cdot)$ and corresponding augmented response $\hat{x}^*(\cdot)$ is

$$- cy^* + [\lambda(t_1)x^*(t_1) - \lambda(t_0)x^*(t_0)] = \int_{t_0}^{t_1} \{[- ck^0(x^*(t),t) + c\nabla k^0(x^*(t),t)x^*(t)] + [- ch^0(u^*(t),t) + \lambda(t)h(u^*(t),t)]\}dt. \tag{6.9}$$

With $x^*(t_0) = x(t_0) = x^0$, the difference (6.9) - (6.8) is given by

$$- c[y^* - y(t_1)] + \lambda(t_1)[x^*(t_1) - x(t_1)] = \int_{t_0}^{t_1} \{ck^0(x(t),t) - ck^0(x^*(t),t) - \nabla(ck^0(x^*(t),t))(x(t) - x^*(t))] + [- ch^0(u^*(t),t) + \lambda(t)h(u^*(t),t) + ch^0(u(t),t) - \lambda(t)h(u(t),t)]\}dt. \tag{6.10}$$

With $c > 0$, and $k^0(\cdot)$ convex in x for every fixed t, it follows from Theorem 2.1 that $ck^0(\cdot)$ is convex so that the first bracket in the integrand is greater than or equal to zero; the second bracket is greater than or equal to zero almost everywhere due to (vii). Hence,

$$- c(y^* - y(t_1)) + \lambda(t_1)[x^*(t_1) - x(t_1)] \geqslant 0. \tag{6.11}$$

If $\theta^1 = \mathbb{R}^n$ or $\theta^1 = \{x^1\}$, then either $\lambda(t_1) = 0$, or $x^*(t_1) - x(t_1) = 0$, and

$$cy^* \leqslant cy(t_1) \tag{6.12}$$

follows in either case.

Now let π be a support hyperplane to θ^1 at $x^*(t_1)$. Since $\lambda^*(t_1)$ is an inward normal at $x^*(t_1)$, and since θ^1 is a closed convex set,

$$\lambda(t_1)(x - x^*(t_1)) \geqslant 0 \tag{6.13}$$

for every $x \in \theta^1$ and one obtains again

$$cy^* \leqslant cy(t_1). \tag{6.14}$$

From the compatibility condition follows

$$c = n(y^*) = a\nabla\phi(y^*), \tag{6.15}$$

where the scalar $a > 0$. Thus,

$$0 \leqslant c(y(t_1) - y^*) = a\nabla\phi(y^*)(y(t_1) - y^*) \tag{6.16}$$

and since $a > 0$,

$$0 \leqslant \nabla\phi(y^*)(y(t_1) - y^*). \tag{6.17}$$

But it follows from hypothesis (iii) that there exists an $F(\cdot)$ such that $\psi(\cdot) = F\circ\phi(\cdot)$ is convex on P. Since $F(\cdot)$ is a strictly increasing function, the derivative $F'(\xi) \geqslant 0 \ \forall\ \xi \in \mathbb{R}$. Hence,

$$\begin{aligned} 0 &\leqslant F'(\phi(y^*))\nabla\phi(y^*)(y(t_1) - y^*) = \\ &\nabla\psi(y^*)(y(t_1) - y^*) \leqslant \psi(y(t_1)) - \psi(y^*), \end{aligned} \tag{6.18}$$

and

$$\psi(y^*) \leqslant \psi(y(t_1)) \Rightarrow y^* \precsim y(t_1) \Rightarrow g(u^*(\cdot)) \precsim g(u(\cdot)) \tag{6.19}$$

for every $u(\cdot) \in \mathcal{F}$. Since $u^*(\cdot) \in \mathcal{F}$, the result is established. ■

Next, a similar approach is used for the programming problem; the theorem is based on results in Mangasarian [7]. The notation $f_I(\cdot)$ simply denotes that subvector of a mapping $f(\cdot)$, whose components are $f_i(\cdot)$, $i \in I$, where I is an appropriate index set.

Theorem 6.2. Let $\precsim$ be a preference relation on P which is a complete preordering of P and let $y^* = g(x^*)$ with $x^* \in X$. Assume:

(i) $S(y^*)$ is such that $n(y^*)$ can be chosen to satisfy $n(y^*) > 0$.

(ii) A utility function $\phi(\cdot)$ is known for $\lesssim$ on P.

(iii) The gradient $\nabla\phi(y^*) > 0$, and $\phi(\cdot)$ satisfies the hypotheses of Theorem 3.3.

(iv) The criterion functions $g_i(\cdot)$, $i = 1, \ldots, N$, are convex with respect to X at $x^* \in X$.

(v) There exist vectors $c \in \mathbb{R}^N$, $\lambda \in \mathbb{R}^m$, $\mu \in \mathbb{R}^k$ such that

$$\begin{aligned} c\nabla g(x^*) + \lambda\nabla f(x^*) + \mu\nabla h(x^*) &= 0, \\ \lambda f(x^*) &= 0, \\ f(x^*) &\leq 0, \\ h(x^*) &= 0, \\ \lambda &\geq 0, \\ c &> 0, \end{aligned} \tag{6.20}$$

where c satisfies the compatibility condition $c = n(y^*)$.

(vi) Let $N = \{1,\ldots,m\}$. With $I = \{i \in N: f_i(x^*) = 0\}$, $f_I(\cdot)$ is differentiable and quasiconvex at x^*.

(vii) The equality constraints $h_i(\cdot)$ are differentiable and both quasiconvex and quasiconcave at x^*.

Then x^* is a preference optimal decision.

Proof: Define $\tilde{f}(\cdot) = (f(\cdot),h(\cdot),-h(\cdot))$ and $\mu = \eta - \nu$ with $(\eta,\nu) \geq 0$. Let $\tilde{\lambda} = (\lambda,\eta,\nu)$. Then the hypothesis (v) may be rewritten as

$$\begin{aligned} c\nabla g(x^*) + \tilde{\lambda}\nabla\tilde{f}(x^*) &= 0, \\ \tilde{\lambda}\tilde{f}(x^*) &= 0, \\ \tilde{f}(x^*) &\leq 0, \\ \tilde{\lambda} &\geq 0. \end{aligned} \tag{6.21}$$

Let $\tilde{N} = \{1,\dots,m+2k\}$ and let $\tilde{I} = \{i \in \tilde{N}: \tilde{f}_i(x^*) = 0\}$, $J = \{i \in \tilde{N}: \tilde{f}_i(x^*) < 0\}$ with $\tilde{I} \cup J = \tilde{N}$. From $\tilde{\lambda} \geqslant 0$, $\tilde{f}(x^*) \leqslant 0$ and $\tilde{\lambda}\tilde{f}(x^*) = 0$, it follows that

$$\tilde{\lambda}_i\tilde{f}_i(x^*) = 0 \quad \forall\, i \in N, \tag{6.22}$$

and consequently that $\lambda_J = 0$. The following arguments hold for every $x \in X$. The general inequality constraint can be written as

$$\tilde{f}_{\tilde{I}}(x) \leqslant 0 = \tilde{f}_{\tilde{I}}(x^*). \tag{6.23}$$

In view of the hypotheses on $f_I(\cdot)$ and $h(\cdot)$, the function $\tilde{f}_{\tilde{I}}(\cdot)$ is quasiconvex at x^*. Thus it follows from Theorem 2.3 that

$$\nabla\tilde{f}_{\tilde{I}}(x^*)(x - x^*) \leqslant 0. \tag{6.24}$$

Furthermore, $\tilde{\lambda}_{\tilde{I}} \geqslant 0$ implies

$$\tilde{\lambda}_{\tilde{I}}\nabla\tilde{f}_{\tilde{I}}(x^*)(x - x^*) \leqslant 0 \tag{6.25}$$

and $\tilde{\lambda}_J = 0$ implies

$$\tilde{\lambda}_J\nabla\tilde{f}_J(x^*)(x - x^*) = 0. \tag{6.26}$$

The addition of the last two equations results in

$$\tilde{\lambda}\nabla\tilde{f}(x^*)(x - x^*) \leqslant 0. \tag{6.27}$$

In view of

$$c\nabla g(x^*) + \tilde{\lambda}\nabla\tilde{f}(x^*) = 0, \tag{6.28}$$

this may also be written as

$$c\nabla g(x^*)(x - x^*) \geqslant 0 \tag{6.29}$$

or

$$\nabla(cg(x^*))(x - x^*) \geqslant 0. \tag{6.30}$$

By hypothesis (iv), along with $c > 0$ and Theorem 2.1, $cg(\cdot)$ is convex at x^*. It then follows from Theorem 2.2 that

$$0 \leq \nabla(cg(x^*))(x - x^*) \leq cg(x) - cg(x^*) \tag{6.31}$$

or

$$cg(x^*) \leq cg(x). \tag{6.32}$$

As in Theorem 6.1, the compatibility condition on c is used to obtain

$$0 \leq \nabla\phi(y^*)(y - y^*), \tag{6.33}$$

and with $F(\cdot)$ and $\psi(\cdot)$ as before,

$$0 \leq F'(\phi(y^*))\nabla\phi(y^*)(y - y^*) = \nabla\psi(y^*)(y - y^*) \leq \psi(y) - \psi(y^*). \tag{6.34}$$

Since $\psi(\cdot)$ is also a utility function for $\precsim$ on P, one has finally $\psi(y) \geq \psi(y^*) \Rightarrow y \succsim y^* \Rightarrow g(x) \succsim g(x^*)$, and since $x^* \in X$, it follows that x^* is preference optimal. ■

The last two theorems are illustrations of a general procedure by means of which *any* sufficiency theorem in optimal control or programming may be used to construct a sufficiency theorem for preference optimality. The procedure is summarized in a programming context; naturally, it is equally applicable for the control problem. For this purpose, let the "standard programming problem" be Problem I with a criterion function $G(\cdot)\colon X \to \mathbb{R}$ replacing the mapping $g(\cdot)$ and with the objective: Minimize $G(x)$ subject to $x \in X$. Let Th be any sufficiency theorem for the "standard programming problem," that is, any theorem which guarantees that a decision x^* satisfies $G(x^*) \leq G(x)\ \forall\ x \in X$. Then the hypotheses of Th, together with suitable assumptions for the preference relation on P, may be used to construct the following meta-sufficiency theorem, i.e., a theorem which is a prescription for the construction of sufficiency theorems for preference optimality.

<u>Theorem 6.3</u>. Let $\precsim$ be a preference relation on P which is a

complete preordering of P and let $y^* = g(x^*)$ with $x^* \in X$. Assume:

(i) It is known that there exists a differentiable utility function $\phi(\cdot)$ for $\precsim$ on P such that $\phi(\cdot)$ is pseudo-convex with respect to Y at $y^* \in Y$ and such that $\nabla\phi(y^*) \neq 0$.

(ii) The indifference sets of $\precsim$ are such that a normal vector $n(y^*)$ to $S(y^*)$ at y^* is known, and has the same orientation as $\nabla\phi(y^*)$.

(iii) All the conditions of Th are satisfied with $n(y^*)g(\cdot)$ replacing $G(\cdot)$.

Then x^* is a preference optimal decision.

Proof: As a consequence of (iii),

$$n(y^*)g(x^*) \leq n(y^*)g(x) \quad \forall\ x \in X.$$

From (ii) it follows that $n(y^*) = a\nabla\phi(y^*)$, $a > 0$, with the result

$$\nabla\phi(y^*)(y - y^*) \geq 0 \quad \forall\ y \in Y.$$

But since $\phi(\cdot)$ is pseudo-convex at y^*, this implies

$$\phi(y^*) \leq \phi(y) \quad \forall\ y \in Y.$$

Consequently,

$$g(x^*) \precsim g(x) \quad \forall\ x \in X,$$

and since $x^* \in X$, x^* is preference optimal. ■

The three fundamental influences in preference optimality are clearly apparent from this theorem, namely, the normal vectors $n(y)$ characterizing the indifference surfaces, the functions $g_i(\cdot)$ characterizing the attainable criteria set, and the utility function $\phi(\cdot)$ characterizing preference. As has already been remarked, properly, all conditions concerning preference optimality should be stated only in terms of the preference relation. In the preceding theorem this would essentially be brought about by conditions on $\lesssim$ guaranteeing the existence of an at least pseudo-convex utility function. With the additional hypothesis of strong convexity of the preference relation, everywhere non-zero Gaussian curvature of the indifference surfaces and compactness, along with convexity of the attainable criteria set, Debreu's Theorem may be used in conjunction with a result, due to Aumann [25] and similar to Theorem 3.3, to assure the existence of a convex utility function $\phi(\cdot)$ for $\lesssim$ on compact subsets of P and hence on $K(t_1)$ or Y. This result, together with Theorem 5.3, allows one to construct sufficiency theorems which require no explicit knowledge of a utility function.

The possible use of the sufficient conditions may become clearer by means of the following simple example. It is cast in a programming framework, since a control example was given earlier. The geometry of the example is such that maximization is more suitable as the basic objective.

To give an indication of the applicability of the present concept, the problem is stated in a biological context. Assume that two different species, 1 and 2, are capable of survival in a given area and that they consume the same type of food. Let x_1 represent the average number of hours that each species has to spend on food foraging per day and x_2 the average amount of time spent by each in the evasion of predators. Let $g_1(x)$ and $g_2(x)$ be the number of the respective species capable of survival under the given conditions x. Now a farmer may control the amount of food and the predators to some extent and he may consider different combinations of these species more desirable than others, thus giving rise in a natural way to a preference relation over the space $g = (g_1, g_2)$.

Let the preference relation be described in the following manner. With $P = \{y \in \mathbb{R}^2 : y >> 0\}$ consider the family of hyperbolas $y_2 = \frac{a}{y_1}$, $a > 0$, $y = (y_1, y_2) \in P$. Let the indifference surfaces of the relation be the members of this family; that is, if $z = (z_1, z_2)$ is any point in P and a is such that z may be written as $z = (z_1, \frac{a}{z_1})$, then

$$S_a(z) \triangleq \{x \in P : x_2 = \frac{a}{x_1}\} \tag{6.36}$$

is the indifference class of z. With

$$\bar{S}_a^+(z) \triangleq \{x \in P : x_2 \geq \frac{a}{x_1}\} \tag{6.37}$$

the statement

$$\forall\, z \in P, \quad x \succsim z \quad \text{iff} \quad x \in \bar{S}_a^+(z),$$

along with an equivalent one for $x \precsim z$, defines a monotone, continuous, and convex preference relation of class C^2 which is a complete preordering of P.† In addition, it follows from Theorem 2.5 that any utility function on P will be quasiconcave, since the set $\{x \in P : x \succsim y\}$ is convex for every $y \in P$.

As has been mentioned among the Assumptions, once an indifference surface of sufficient differentiability is known, one also has knowledge of a normal vector field to this surface. In order to apply the given

† One should say here, of "at least" class C^2.

conditions a normal vector must be known for the necessity part and the existence of a suitable utility function must be assured for the sufficient part. For every $y \in P$, $a > 0$, the following choice is made for these:

$$n(y) = \left(\frac{a}{y_1^2}, 1\right) \text{ and } \phi(y) = y_1 y_2. \tag{6.38}$$

This takes care of the preference on P.

With maximization as the basic objective, consider then the following specific problem: Obtain preference optimal decisions for

$$g_1(x) = \sqrt{x_1 x_2} \quad \text{and} \quad g_2(x) = x_1 - x_2 \tag{6.39}$$

subject to the inequality constraints

$$f_1(x) = x_2 - x_1 < 0, \quad f_2(x) = x_1 - 1 \leq 0, \quad f_3(x) = -x_2 < 0. \tag{6.40}$$

The sets X and Y are defined by

$$X = \{x \in \mathbb{R}^2 : f(x) < 0\} \tag{6.41}$$

and

$$Y = g[X]. \tag{6.42}$$

These sets are sketched in Figure 5.

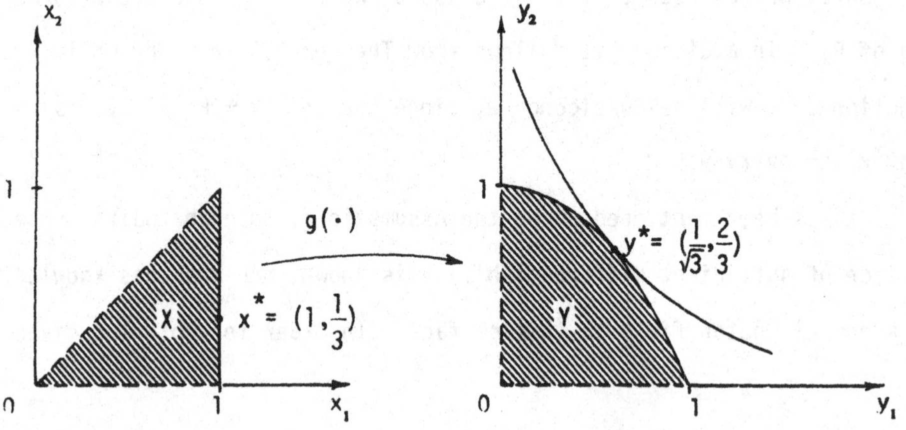

Fig. 5. Functional constraint set X, attainable criteria set Y, together with preference optimal decision and criterion value.

Solution: The solution consists of two parts, application of the necessary condition Theorem 5.3 to obtain a candidate (or candidates) and a check of sufficiency conditions to assure preference optimality.

Part I. With three inequality constraints, the necessary conditions embodied in Theorem 5.3 require

$$-\left(\frac{1}{2}c_1\sqrt{\frac{x_2}{x_1}} + c_2, \frac{1}{2}c_1\sqrt{\frac{x_1}{x_2}} - c_2\right) + (-\lambda_1 + \lambda_2, \lambda_1 - \lambda_3) = 0. \tag{6.43}$$

However, $f_1(\cdot)$ and $f_3(\cdot)$ are strict inequalities for every $x \in \mathbb{R}^2$, hence never active, so that $\lambda_1 = \lambda_3 = 0$. An interior point of X cannot be a solution so that only the possibility with $f_2(\cdot)$ active remains. With $f_2(\cdot)$ as the active constraint, $x_1 = 1$, $\lambda_2 \neq 0$ and

$$\begin{aligned} -\frac{1}{2}c_1\sqrt{x_2} - c_2 &= -\lambda_2, \\ -\frac{1}{2}c_1\frac{1}{\sqrt{x_2}} + c_2 &= 0, \end{aligned} \tag{6.44}$$

are all that remains of the necessary conditions. The solution of the system yields $x_2 = \frac{1}{4}\left(\frac{c_1}{c_2}\right)^2$.

To apply the compatibility condition, y is needed in terms of c: that is,

$$\begin{aligned} y_1 &= \frac{c_1}{2c_2}, \\ y_2 &= 1 - \left(\frac{c_1}{2c_2}\right)^2. \end{aligned} \tag{6.45}$$

Since (y_1, y_2) must belong to $S_a(y)$, $a = \left(\frac{c_1}{2c_2}\right)\left[1 - \left(\frac{c_1}{2c_2}\right)^2\right]$, and with the above choice of normal, n(y) in terms of c becomes

$$\tilde{n}(c) = (\frac{2c_2}{c_1}[1 - (\frac{c_1}{2c_2})^2], 1). \quad (6.46)$$

Thus,

$$c_1 = \frac{2c_2}{c_1}[1 - (\frac{c_1}{2c_2})^2], \quad c_2 = 1, \quad (6.47)$$

with the result

$$c_1 = \frac{2}{\sqrt{3}}, \quad c_2 = 1, \quad (6.48)$$

along with $x_2 = \frac{1}{3}$. Note that the specific values for the c_i are illusory, since all the results remain unchanged with $c_1 = \frac{2\delta}{\sqrt{3}}$, $c_2 = \delta$, $\delta > 0$.

Part II. Claim: $x^* = (1, \frac{1}{3})$ is a preference optimal decision with $y^* = (\frac{1}{\sqrt{3}}, \frac{2}{3})$.

Note first that $\phi(\cdot)$ is not concave. Essentially, this leaves two approaches, either to show that $\phi(\cdot)$ is pseudo-concave and to make implicit use of Theorem 6.3 or to show that $\phi(\cdot)$ satisfies the hypotheses of Theorem 6.2, and in particular, those which concern Theorem 3.3.

The function $\phi(\cdot)$ is indeed pseudo-concave, since $y \in P$, $\nabla\phi(y^*)(y - y^*) \leq 0$ implies $\phi(y) \leq \phi(y^*)$; i.e., $y_1 y_2 \leq y_1^* y_2^*$ as long as $y_2 \leq -(\frac{y_2^*}{y_1^*})y_1 + 2y_2^*$. This is most easily seen from the fact that $y_1^* y_2^*$ is the largest possible area of an inscribed rectangle.

However, it may usually be quite difficult to show pseudo-concavity or concavity directly, so that it is instructive to illustrate the application of Theorem 6.2. The satisfaction of the hypotheses of Theorem 3.3 is of particular interest, since all other conditions are obviously satisfied. Naturally, the conditions (i) - (v) of the theorem are modified to conform to the basic objective of maximization.

(i) Since $\nabla\phi(y) = (y_2, y_1) >> 0 \ \forall \ y \in P$, $\phi(\cdot)$ has neither a stationary nor a maximum point on P.

(ii) The quadratic form $Q(\cdot)$ is given by

$$Q(\xi,\xi) = (\xi_1,\xi_2)\begin{bmatrix} \phi_{11}(y) + \sigma(y)\phi_1^2(y) & \phi_{12}(y) + \sigma(y)\phi_1(y)\phi_2(y) \\ \phi_{21}(y) + \sigma(y)\phi_1(y)\phi_2(y) & \phi_{22}(y) + \sigma(y)\phi_2^2(y) \end{bmatrix}\begin{pmatrix} \xi_1 \\ \xi_2 \end{pmatrix} \tag{6.49}$$

with characteristic polynomial

$$\Gamma_Q(\alpha) = \alpha^2 - \alpha\sigma(y)(y_1^2 + y_2^2) - (1 + 2\sigma(y)y_1y_2), \tag{6.50}$$

where

$$\begin{aligned} &\phi_1(y) = y_2, \quad \phi_2(y) = y_1, \quad \phi_{12}(y) = \phi_{21}(y) = 1, \\ &\text{and} \quad \phi_{11}(y) = \phi_{22}(y) = 0 \end{aligned} \tag{6.51}$$

have been used. A choice of

$$\sigma(y) = -\frac{1}{2y_1y_2} \tag{6.52}$$

yields

$$\begin{aligned} \alpha_1 &= 0 \\ \alpha_2 &= -\frac{y_1^2 + y_2^2}{2y_1y_2} \end{aligned} \tag{6.53}$$

assuring the negative semidefiniteness of $Q(\cdot)$ (by an extension of Theorem 2.7). It is possible to use Theorem 3.3 in a constructive manner, for with

$$\sigma(y) = \frac{F''(\phi(y))}{F'(\phi(y))} \tag{6.54}$$

one may obtain the desired transformation $F(\cdot)$ from

$$F''(\phi) + \frac{1}{2\phi}F'(\phi) = 0 \tag{6.55}$$

as

$$F(\phi) = \frac{1}{2}d_1\sqrt{\phi} + d_2. \tag{6.56}$$

The d_i are irrelevant and a concave utility function $\psi(\cdot)$ is defined by

$$\psi(y) = F\circ\phi(y) = \sqrt{y_1 y_2}. \tag{6.57}$$

(iii) The matrix associated with $Q(\xi,\xi)$ has rank 1, and $[\phi_{ij}(y)]$ has rank 2 for every $y \in P$.

(iv) Here

$$\Gamma(\alpha) = \alpha^2 - 1 \tag{6.58}$$

and

$$\Gamma^*(\alpha) = -\frac{2y_1y_2}{y_1^2 + y_2^2} - \alpha \tag{6.59}$$

so that

$$\begin{aligned} &D_0 = 1, \quad D_1 = 0, \quad D_2 = -1; \\ &D_0^* = 1, \quad D_1^* = -\frac{2y_1y_2}{y_1^2 + y_2^2}. \end{aligned} \tag{6.60}$$

Since maximization is the objective,

$$G(t) = \sup\{\frac{D}{k^2(y)D_1^*}: \phi(y) = t\} = \frac{1}{2t} < \infty \tag{6.61}$$

for every fixed $t \in (\varepsilon,\beta) = (0,\infty)$.

(v) Let a minorant function $\gamma(\cdot): [\varepsilon,\beta) \to \mathbb{R}$ be defined by

$$\gamma(t) = \frac{H'(t)}{H(t)}. \tag{6.62}$$

An obvious choice for $H(\cdot)$ is $H(t) = \sqrt{t}$. Then,

$$G(t) = \frac{1}{2t} \geq \frac{1}{2t} \tag{6.63}$$

follows.

Hence, $x^* = (1,\frac{1}{3})$ is a preference optimal decision with resultant criterion value $y^* = (\frac{1}{\sqrt{3}},\frac{2}{3})$.

In conclusion then, if x_1 and x_2 are given in hours and y_1, y_2 in terms of hundreds of members of a species, the answer for the farmer would be: Allow the two species 1 hour for feeding, chase them about for another $\frac{1}{3}$ of an hour, and as a result, approximately 58 of the first species and 67 of the second species will survive.

7. A COMPARISON WITH OTHER OPTIMALITY CONCEPTS

In this section, a brief comparison with other optimality concepts in vector-valued criteria problems is given.

Instead of taking the preference relation as the primitive notion of the theory, one could have used utility as the primitive;† that is, one could have introduced a differentiable utility function to provide a one-dimensional ordinal comparison of the criteria. This would change none of the results. It must be emphasized that in view of its nonuniqueness such a function does no more than to provide an ordering of the choices;

† A primitive is a concept which requires no explanation in terms of other more basic concepts; that is, there are none which are more basic.

it does not assign numerical amounts to physical quantities. Thus, it is meaningless to compare, that is, to add or subtract, the numerical values of utility functions.†

On the other hand, it is possible to introduce one-dimensional comparison functions of the criteria whose numerical values may have physical interpretation. For example, assume that two individuals, J_1 and J_2, wish to travel together and that in one day they wish to get as far away from a given town 0 as possible. Because of their separate future destinations, J_1 would prefer to travel due east and J_2 due north. In view of their desire to travel together they form a compromise concerning their destination at the end of the day in that J_1 insists only on an eastern component of

$$g_1(x) = x_2 + \cos x_1 \tag{7.1}$$

in tens of miles, and J_2 insists on a northern component of

$$g_2(x) = 2x_2 \sin x_1 \tag{7.2}$$

in tens of miles. Here, $x_1 \in [0,\frac{\pi}{2}]$ is descriptive of their directional bias and $x_2 \in [0,1]$ is representative of their food provision for the day. Note that x_1 is not the polar angle θ, and that x_2 is not the food per day; these variables only describe the effect of these quantities on the components of the destination at the end of the day. For example, if their destination at the end of the day is to be $y = (1,2)$, then this corresponds to $x = (\frac{\pi}{2},1)$. Thus,

$$g_i(\cdot): X \to \mathbb{R}, \quad i = 1,2,$$

† Additive utility theory is a partial exception to this statement [8].

where $X = \{x \in \mathbb{R}^2: x_1 \in [0,\frac{\pi}{2}], x_2 \in [0,1]\}$. In view of their agreed upon restrictions concerning their common possible destinations y at the end of the day, this y must be within the set

$$Y = \{y \in \mathbb{R}^2: 0 \leq y_2 \leq 2y_1, (y_1 - 1)^2 + (\frac{y_2}{2})^2 \leq 1\}. \tag{7.3}$$

The maximum distance they can cover during that day is obtained by maximizing the objective function

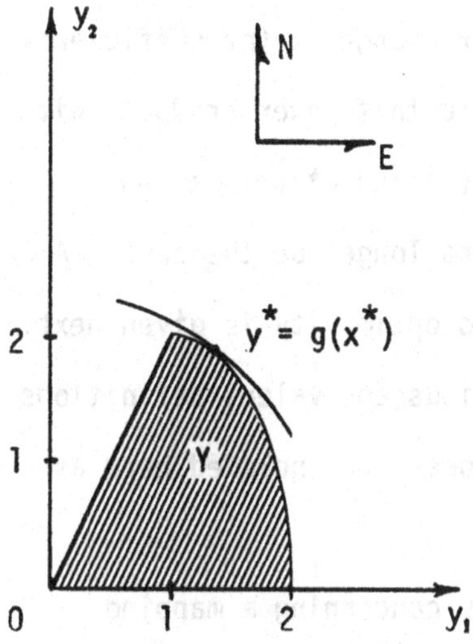

Fig. 6. The optimal compromise destination of J_1 and J_2.

$$R\circ g(x) = [\sum_{i=1}^{2} g_i^2(x)]^{1/2} \tag{7.4}$$

subject to $x \in X$, where $R(\cdot): Y \to \mathbb{R}$, and where the value $R\circ g(x)$ is the actual distance in tens of miles that they are capable of travelling that day. Thus, if $R\circ g(x^*) = \max \{R\circ g(x): x \in X\}$, then $y^* = g(x^*)$ is their optimal destination at the end of the day.

More generally, let $g(\cdot): X \to Y$ and let

$$\bar{y}_i = g_i(\bar{x}^i) = \min \{g_i(x): x \in X\}, \quad i = 1, \ldots, N. \tag{7.5}$$

Define regret functions $R(\cdot): Y \to \mathbb{R}$ by

$$R(y) = [\sum_{i=1}^{N} (\bar{y}_i - y_i)^p]^{1/p}. \tag{7.6}$$

Then Theorem 5.3 applies as a necessary condition for the minimization of $R\circ g(x)$ subject to $x \in X$, as long as the $g_i(\cdot)$ are sufficiently differentiable. Additional detail concerning such regret functions $R(\cdot)$ and some of their properties may be found in Yu and Leitmann [1]. Further treatment of one-dimensional comparison functions with physical meaning, along with necessary conditions for their minimization are given by Salama and Gourishankar [26].

When such one-dimensional comparison functions are used, the necessary conditions as given here are altered only in that it generally is no longer possible to assert $c > 0$ in the maximum principle, and that, in particular, $c^* = 0$ is a possibility. The major change in the sufficiency conditions concerns the assumptions which assure that inner products with the vector c^* are convex functions, e.g., $c^*k^0(\cdot)$ and $c^*g(\cdot)$ may no longer be convex functions, since $c^* > 0$ need no longer be the case. A comparison of preference optimality with Pareto-optimality is given next.

In the treatment of Pareto-optimality various equivalent definitions are in use. For convenience, the more common ones are included here as a lemma.

Lemma 7.1. The following three statements concerning a mapping $u^*(\cdot) \in \mathcal{F}$ and a corresponding criterion vector $g(u^*(\cdot))$ imply each other:

(i) For every $u(\cdot) \in \mathcal{F}$ such that $g(u(\cdot))$ is comparable to $g(u^*(\cdot))$, $g(u(\cdot)) \leq g(u^*(\cdot)) \Rightarrow g(u(\cdot)) = g(u^*(\cdot))$.

(ii) $\nexists\, u(\cdot) \in \mathcal{F}$ such that $g(u(\cdot)) < g(u^*(\cdot))$.

(iii) For every $u(\cdot) \in \mathcal{F}$, either $g(u(\cdot)) = g(u^*(\cdot))$ or there exists at least one $j \in \{1,\ldots,N\}$ such that $g_j(u(\cdot)) > g_j(u^*(\cdot))$. □

The proof is simple and is omitted. Thus, a Pareto optimal control $u^*(\cdot)$ is nothing but a control which results in a minimal element $y^* \in K(t_1) \subset P$ for the partial ordering given by the coordinatewise ordering of P. Clearly, then, any of the above statements may be taken as the definition of Pareto optimality.

Conditions are now given subject to which a preference optimal control is also Pareto-optimal. The conditions are given for the control problem; of course, they are equally applicable to the programming problem.

Theorem 7.1. Let $\precsim$ be monotone and a complete preordering of $K(t_1) \subset \mathbb{R}^N$. Assume that $u^*(\cdot) \in \mathcal{T}$ is a preference optimal control. Then $u^*(\cdot)$ is also a Pareto optimal control.

Proof: Take (ii) of Lemma 7.1 as the definition of Pareto-optimality. Let $u^*(\cdot) \in \mathcal{T}$ be preference optimal, that is, $g(u^*(\cdot)) \precsim g(u(\cdot))$ $\forall\, u(\cdot) \in \mathcal{T}$. Assume that $u^*(\cdot)$ is not Pareto-optimal; then there exists a $\bar{u}(\cdot) \in \mathcal{T}$ such that $g(\bar{u}(\cdot)) < g(u^*(\cdot))$. But since $\precsim$ is a monotone preference relation, this implies $g(\bar{u}(\cdot)) \prec g(u^*(\cdot))$, which is a contradiction to the preference optimality of $u^*(\cdot)$. Hence $u^*(\cdot)$ is Pareto-optimal. ■

8. CONCLUSIONS AND SOME POSSIBLE GENERALIZATIONS

An optimal concept for problems with a vector-valued criterion has been presented. It is based on ordering the criteria space by means of a complete preordering $\precsim$, and on the subsequent definition of preference optimality in terms of a least element for $\precsim$. The derivation of necessary conditions involved assumptions on $\precsim$ which guaranteed the existence of a C^2-utility function $\phi(\cdot)$ for $\precsim$; it was shown that a maximum principle involving a scalar combination $cg(\cdot)$ of the criteria was one such condition, another being a compatibility condition of the form

$n(y^*(c^*;t_1)) = c^*$. It must again be emphasized that the necessary conditions may be applied without any explicit knowledge of a utility function.

Sufficient conditions, however, may be based on the assured existence of a suitable utility function or its explicit knowledge.

It would seem that the following generalizations and refinements of the theory are desirable:

(i) All theorems are stated with $\precsim$ as a complete preodering. This was done to conform with the definition of preference optimality which was given in terms of least elements for a complete preordering. Some of the results are applicable for binary relations with less structure, as long as a statement such as "least preferred" retains meaning. Thus the meaning of "least preferred" should be extended beyond minimal and least elements for partial orderings; furthermore, the required structure on $\precsim$ should be eased, in particular, the differentiability assumption. Naturally, this should be accomplished in such a way that one may still derive most of the preceding conditions, especially the necessary conditions in terms of a maximum principle.

(ii) The corresponding existence theorem should be broadened.

(iii) Additional sufficient conditions which make no use of a utility function should be derived.

(iv) A numerical algorithm for the calculation of the c_i, which makes use only of the indifference curves, would be useful.

(v) Additional comparisons to other optimality concepts in multi-criteria problems should be made.

PART II

APPLICATIONS OF PARETO-OPTIMALITY IN MECHANICS

1. INTRODUCTION

Vector-valued criteria have long been used in Economics and in business applications, along with game theory. However, little has been done along this line in Mechanics or for that matter in other engineering fields. The optimality concept which is to be used here is that of Pareto-optimality.

The first application concerns the introduction of measuring devices into an optimally controlled system. Naturally, one would like to measure or observe the undisturbed system. This may be possible in theory, but it is not possible in practice. In any experiment or operation the interaction of system, instrumentation, and investigator is observed. Thus, any measurement constitutes a disturbance of the system. The objective, then, is to minimize the disturbance to the system, while at the same time maintaining optimal control. Only one example is given here; several other examples may be found in Leitmann and Stadler [27].

The second set of applications deals with proposed "natural" shapes. The use of the word "natural" is prompted by the resemblance of some of the results to structural shapes found in nature (one needs a little imagination here). Specifically, when the purpose of the structure has been defined, that is, the loading which it is to support and the geometric constraints to which it is to be subjected, then the optimal (natural) structure (defined in terms of shape, loading, and material parameters, etc.) to fulfill this purpose is one which results in a Pareto-optimal balance between the bulk of the structure and the total stored energy corresponding to this choice of shape and material parameters. Again, only a single simple example is given here; several other examples, along with general properties of such structures, may be found in Stadler [28].

2. OPTIMAL PLACEMENT OF A MEASURING DEVICE

Considered here is an optimally driven disc upon which is to be placed a measuring device, say to measure the instantaneous angular acceleration. The problem then is where to place the device so as to disturb the system "least." Essentially, two separate subproblems are worked. The first is an optimal control problem to obtain a control which strikes a Pareto balance between the two criteria, the other a programming problem to calculate a Pareto-optimal placement of the device.

In Leitmann and Stadler [27] an additional problem is considered: the choice between two different measuring devices, different in the sense that they disturb different parts of the system.

2.1. The "Isolated System". A disc rotates with an angular speed z^0 about a fixed axis through its center 0. The moment of inertia of the disc about this axis is I. Its angular speed at any time $t \in [0,1]$ is denoted by $z(t)$, where $z(\cdot): [0,1] \to \mathbb{R}$. The disc is to be brought to rest by a Lebesgue measurable torque $u(\cdot): [0,1] \to \mathbb{R}$, with $|u| < \infty$, in such a manner that $\int_0^1 u^2(t)dt$ is a minimum. The optimal control problem is formulated by incorporating the desired terminal state $z(1) = 0$ (rest) in the criterion by means of a weighting constant k.† (See Figure 1.) The objective then is to minimize

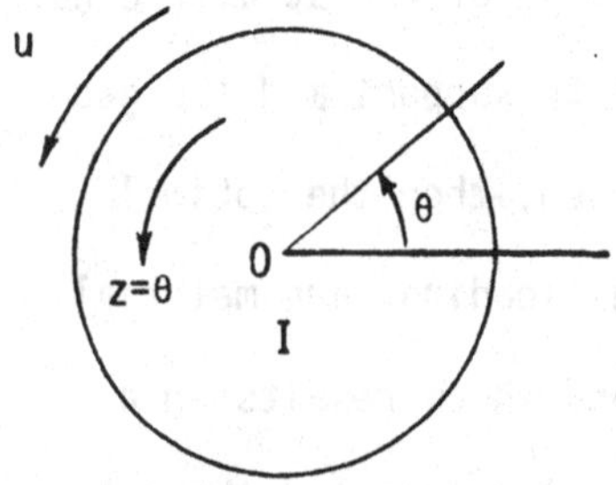

Fig. 1. The "Isolated System".

$$g_1(u(\cdot)) = \frac{1}{2}kz^2(1) + \frac{1}{2}\int_0^1 u^2(t)dt \tag{2.1}$$

subject to Euler's second law (the law of angular momentum)††

$$\dot{z} = \frac{u}{I}, \quad z(0) = z^0 \tag{2.2}$$

with u unconstrained and with $k > 0$. The optimal control (linear equations with quadratic criterion [30])

† In essence the use of a weighted sum of the criteria $\int_0^1 u^2(t)dt$ and z(1) already constitute a problem in Pareto-optimization.

†† For a basic, well written discussion of Mechanics see Fox [29].

$$\hat{u}(t) = -\frac{kz^0 I}{I^2 + k} \tag{2.3}$$

with corresponding solution $\hat{z}(\cdot)$ of the state equation (2.2) defined by

$$\hat{z}(t) = \frac{z^0}{I^2 + k}[I^2 + k(1-t)], \tag{2.4}$$

and with criterion value

$$g_1(\hat{u}(\cdot)) = \frac{k(z^0 I)^2}{2(I^2 + k)} \cdot \tag{2.5}$$

2.2. The "Disturbed" System. A measuring device of mass m, say an accelerometer, now is to be attached to the disc in such a manner that the system is disturbed as little as possible; that is, one wishes to come close to measuring the acceleration of the isolated system. Thus, the device is to be attached in such a way that both the criterion $g_1(u(\cdot))$, and in addition, the mean-square error between the speed $\hat{z}(t)$ of the isolated system and that of the disturbed system are optimized in the sense of Pareto-optimality.

The disturbance to the system is characterized by allowing the moment of inertia to be a function $x_1(\cdot)$: $[0,1] \to \mathbb{R}$ of time. With angular speed $x_2(\cdot)$: $[0,1] \to \mathbb{R}$, torque $u_1(\cdot)$: $[0,1] \to \mathbb{R}$ and rate of change of moment of inertia $u_2(\cdot)$: $[0,1] \to \mathbb{R}$, the new vector-valued criteria problem may be formulated as:

Obtain Pareto-optimal controls $u^*(\cdot)$ for†

$$g_1(u(\cdot)) = \int_0^1 (kx_2(t)\frac{u_1(t) - x_2(t)u_2(t)}{x_1(t)} + \frac{1}{2}u_1^2(t))dt, \tag{2.6}$$

† $g_1(u(\cdot))$ may be written in this form because the initial manifold consists of $\theta^0 = \{z^0\}$.

$$g_2(u(\cdot)) = \frac{1}{2}\int_0^1 (\hat{z}(t) - x_2(t))^2 dt, \tag{2.7}$$

subject to

$$\begin{aligned} \dot{x}_1 &= u_2 \\ \dot{x}_2 &= \frac{u_1 - x_2u_2}{x_1}, \quad x_2(0) = z^0, \quad x_2(1) \text{ arbitrary.} \end{aligned} \tag{2.8}$$

The control constraint set U, and the initial and final moment of inertia $x_1(0)$ and $x_1(1)$ have been left unspecified so that a number of different possibilities may be considered; also, the state constraint $x_1(t) \geqslant I$ $\forall\ t \in [0,1]$ is omitted for the moment.

The application of necessary conditions for Pareto-optimality as given in Yu and Leitmann [24] quite generally requires an H-function

$$\begin{aligned} H(\lambda,x,u) = &- c_1(kx_2\frac{u_1 - u_2x_2}{x_1} + \frac{1}{2}u_1^2) - \frac{1}{2}c_2(\hat{z}(t) - x_2)^2 + \\ &\lambda_1u_2 + \lambda_2\frac{u_1 - u_2x_2}{x_1} \end{aligned} \tag{2.9}$$

and adjoint equations

$$\dot{\lambda}_1 = - c_1kx_2\frac{u_1 - u_2x_2}{x_1^2} + \lambda_2\frac{u_1 - u_2x_2}{x_1^2}, \tag{2.10}$$

$$\dot{\lambda}_2 = c_1k\frac{u_1}{x_1} - c_2(\hat{z} - x_2) + (\lambda_2 - c_1kx_2)\frac{u_2}{x_1} - c_1kx_2\frac{u_2}{x_1}. \tag{2.11}$$

In order to proceed any further some special cases need be considered. Only c_1, $c_2 > 0$ will be considered here; c_1 and c_2 equal to zero were treated in Leitmann and Stadler [27].

2.2.1. The Control Constraint Set $U = \mathbb{R}^2$. As before, the notation $U = \mathbb{R}^2$ simply means that arbitrary bounded and Lebesgue measurable controls with values in $\mathbb{R}^2$ are to be considered. In view of this arbitrariness one must have

$$\frac{\partial H}{\partial u_1} = 0 = - c_1 k\frac{x_2}{x_1} - c_1 u_1 + \frac{\lambda_2}{x_1} \Rightarrow u_1(t) = \frac{1}{c_1 x_1(t)}(\lambda_2(t) - c_1 k x_2(t)) \tag{2.12}$$

and

$$\frac{\partial H}{\partial u_2} = 0 = c_1 k\frac{x_2^2}{x_1} + \lambda_1 - \lambda_2\frac{x_2}{x_1} \Rightarrow \lambda_1(t) = \frac{x_2(t)}{x_1(t)}(\lambda_2(t) - c_1 k x_2(t)). \tag{2.13}$$

For convenience the argument t is omitted in the following manipulations. The substitution of equation (2.13) in the adjoint equation (2.10) yields

$$\dot{\lambda}_1 = \frac{\dot{x}_2}{x_2}\lambda_1, \tag{2.14}$$

and the substitution of equation (2.12) into the adjoint equation (2.11) leads to

$$\dot{\lambda}_2 = (\frac{k}{x_1^2} + \frac{\dot{x}_1}{x_1})(\lambda_2 - c_1 k x_2) - c_2(\hat{z} - x_2) - c_1 k x_2\frac{\dot{x}_1}{x_1}. \tag{2.15}$$

Equation (2.13) together with its derivative and equation (2.14) result in

$$\dot{\lambda}_2 = \frac{\dot{x}_1}{x_1}(\lambda_2 - c_1 k x_2) + c_1 k\dot{x}_2. \tag{2.16}$$

The comparison of equation (2.16) with (2.15) along with

$$\frac{d}{dt}(x_1 x_2) = u_1 = \frac{1}{c_1 x_1}(\lambda_2 - c_1 k x_2) \tag{2.17}$$

then provides

$$x_2(t) = \hat{z}(t). \tag{2.18}$$

If $z^0 = 0$ there is nothing to show. Hence, assume $z^0 \neq 0$. Then $\hat{z}(t) \neq 0 \ \forall \ t \in [0,1]$ and equation (2.14) may be integrated to obtain

$$\lambda_1(t) = \frac{\lambda_1(0)}{z^0}\hat{z}(t) \tag{2.19}$$

along with

$$\lambda_1(0)x_1(t) = z^0(\lambda_2(t) - c_1k\hat{z}(t)) \tag{2.20}$$

from expression (2.13). The corresponding control is

$$u_1(t) = \frac{\lambda_1(0)}{c_1z^0}\ . \tag{2.21}$$

These last three conditions must be satisfied no matter what the values of $x_1(0)$ and $x_1(1)$ are. Furthermore, since $x_2(1)$ is arbitrary, $\lambda_2(1) = 0$ is necessary. Next, the possible combinations of initial and final moment of inertia are considered.

Case 1. $x_1(0)$ arbitrary, $x_1(1)$ fixed or arbitrary. Since $x_1(0)$ is arbitrary, $\lambda_1(0) = 0 \Rightarrow 0 = \lambda_2(t) - c_1k\hat{z}(t) \Rightarrow \lambda_2(1) = c_1k\hat{z}(1) = c_1kz^0(\frac{I^2}{I^2 + k}) = 0$ iff $z^0 = 0$.

Case 2. $x_1(0) = v_0$, $x_1(1)$ arbitrary. Since $x_1(1)$ is arbitrary, $\lambda_1(1) = 0 \Rightarrow \lambda_1(1) = \lambda_1(0)(\frac{I^2}{I^2 + k}) = 0$ iff $\lambda_1(0) = 0$ in view of $I^2 \neq 0$. The rest then follows as in Case 1.

Case 3. $x_1(0) = v_0$, $x_1(1) = v_1$. This case was not considered in Leitmann and Stadler [27]. Since $x_1(0)$, $x_1(1)$ are given, $\lambda_1(0)$ and $\lambda_1(1)$ are abitrary. With

$$u_1(t) = \frac{\lambda_1(0)}{c_1z^0} \tag{2.22}$$

one may solve

$$\frac{d}{dt}(x_1(t)\hat{z}(t)) = \frac{\lambda_1(0)}{c_1z^0} \tag{2.23}$$

to obtain

$$x_1(t) = \frac{1}{\hat{z}(t)}\left[\frac{\lambda_1(0)}{c_1 z^0}t + v_0 z^0\right]. \tag{2.24}$$

The constant $\lambda_1(0)$ now is adjusted to satisfy $x_1(1) = v_1$, resulting in

$$\lambda_1(0) = \frac{c_1(z^0)^2}{I^2 + k}[v_1 I^2 - v_0(I^2 + k)]. \tag{2.25}$$

The remaining condition to be satisfied is equation (2.20) evaluated at $t = 1$. With $\lambda_2(1) = 0$ this is possible only if v_0 and v_1 are related by

$$v_0 = \frac{I^2}{I^2 + k}\,\frac{v_1^2 + k}{v_1}\,. \tag{2.26}$$

Thus, no solution exists for arbitrarily specified initial and final moments of inertia. However, if one chooses them as related by equation (2.26) and requires that $v_1 \geqslant I$, and that $k \leqslant v_1 I$, then the control

$$u^*(t) = \frac{kI^2}{v_1}\left(-\frac{z^0}{I^2 + k}, \frac{v_1^2 - I^2}{[I^2 + k(1-t)]^2}\right) \tag{2.27}$$

results in

$$x_1^*(t) = \frac{I^2}{v_1}\,\frac{v_1^2 + k(1-t)}{I^2 + k(1-t)} \tag{2.28}$$

with $x_1^*(t) \geqslant I \;\forall\; t \in [0,1]$, and

$$x_2^*(t) = \hat{z}(t). \tag{2.29}$$

Furthermore, this control is Pareto-optimal since it results in $g_2(u^*(\cdot)) = 0$, the absolute minimum of $g_2(u(\cdot))$.

In conclusion, then, there generally does not exist a control $u^*(\cdot)$, with $u \in \mathbb{R}^2$, unless $z^0 = 0$. However, if one allows for a movable device whose initial and final position are implicitly restricted by (2.26) above, then one may achieve zero error by suitably moving the device, and by using an external torque $u^*(\cdot)$ which is related to $\hat{u}(\cdot)$ by

$$u_1^*(t) = \frac{I}{v_1}\hat{u}(t). \tag{2.30}$$

Note that $v_1 = I \Rightarrow v_0 = I \Rightarrow x_1^*(t) = I$; that is, placing the device at the center of the disc causes no disturbance to the system, but in most cases, neither does it measure anything.

2.2.2. The Control Constraint Set $U = \mathbb{R}^1$. The device is now placed on the disc at an arbitrary but fixed location, a more realistic situation, since a measuring device usually is not moved about during an experiment. In this case one might just as well consider a problem with an arbitrary, Lebesgue measurable and bounded control. The problem is: Obtain Pareto-optimal control $u^*(\cdot)$ for

$$\begin{aligned} g_1(u(\cdot)) &= \int_0^1 \left(kx(t)\frac{u(t)}{v} + \frac{1}{2}u^2(t)\right)dt \\ g_2(u(\cdot)) &= \frac{1}{2}\int_0^1 (\hat{z}(t) - x(t))^2 dt \end{aligned} \tag{2.31}$$

subject to

$$\dot{x} = \frac{u}{v} \quad \text{with} \quad x(0) = z^0. \tag{2.32}$$

Here $v = I + mr^2$, where r is the distance of the device from 0. The H-function is

$$H(\lambda,x,u) = -c_1\left(kx\frac{u}{v} + \frac{1}{2}u^2\right) - \frac{1}{2}c_2(\hat{z}(t) - x)^2 + \lambda\frac{u}{v} \tag{2.33}$$

with corresponding adjoint equation

$$\dot{\lambda} = c_1 k\frac{u}{v} - c_2(\hat{z}(t) - x). \tag{2.34}$$

The condition

$$\frac{\partial H}{\partial u} = 0 = - c_1 kx\frac{1}{v} - c_1 u + \lambda\frac{1}{v} \tag{2.35}$$

results in

$$u(t) = \frac{1}{c_1 v}(\lambda(t) - c_1 kx(t)). \tag{2.36}$$

With this $u(\cdot)$ the state equation becomes

$$\dot{x}(t) = \frac{1}{c_1 v^2}(\lambda(t) - c_1 kx(t)). \tag{2.37}$$

Differentiation of this equation and the subsequent substitution of the adjoint equation yields

$$\ddot{x}(t) - \Omega^2 x(t) = - \Omega^2 \hat{z}(t), \quad \Omega^2 = \frac{c_2}{c_1 v^2}. \tag{2.38}$$

With $\lambda(1) = 0$ and $x(0) = z^0$, the solution $x^*(\cdot)$ is defined by

$$x^*(t) = (a - \frac{kx^*}{v^2})\frac{\sinh \Omega t}{\Omega \cosh \Omega} + \hat{z}(t) \tag{2.39}$$

corresponding to a control $u^*(\cdot)$ which satisfies

$$u^*(t) = - av + (av - \frac{kx^*}{v})\frac{\cosh \Omega t}{\cosh \Omega}, \tag{2.40}$$

where

$$x^* = x^*(1) = v^2\frac{a \tanh \Omega + \Omega(z^0 - a)}{v^2\Omega + k \tanh \Omega} \quad \text{and} \quad a = \frac{kz^0}{I^2 + k}. \tag{2.41}$$

To confirm Pareto-optimality one may reason as follows: This is the unique optimal control for the scalar criterion

$$J(u(\cdot)) = c_1 g_1(u(\cdot)) + c_2 g_2(u(\cdot)), \quad (c_1,c_2) >> 0; \tag{2.42}$$

hus, $J(u(\cdot)) - J(u^*(\cdot)) > 0$ for every admissible $u(\cdot)$, so that at least one $\Delta g_i(u(\cdot)) = g_i(u(\cdot)) - g_i(u^*(\cdot)) > 0$ for every such control.†

The corresponding criterion values are

$$g_1(u^*(\cdot)) = \frac{1}{2}k(x^*)^2 + \frac{c_2}{4\Omega^2 c_1}\{(a - \frac{kc_1}{c_2}x^*\Omega^2)^2(\frac{\tanh\Omega}{\Omega} + \mathrm{sech}^2\Omega) -$$

$$4a(a - \frac{c_1 k}{c_2}x^*\Omega^2)\frac{\tanh\Omega}{\Omega} + 2a^2\}, \tag{2.43}$$

$$g_2(u^*(\cdot)) = \frac{1}{4\Omega^2}(a - \frac{c_1 k}{c_2}x^*\Omega^2)^2(\frac{\tanh\Omega}{\Omega} - \mathrm{sech}^2\Omega).$$

Recall that the value of the parameter Ω depends on the placement of the device and on the emphasis which is to be placed on the criteria. With this in mind, an additional problem may now be formulated to determine the "best" placement of the device.

2.3. The "Best" Constant Placement. In the following multicriteria programming problem let $x_1 = \Omega$, $x_2 = x^* = x^*(1)$ and let the range of placement of the device be given by $d \leq r \leq e$ with corresponding bounds of moment of inertia $v_0 \leq v \leq v_1$. For a weighting c of the criteria and with

$$f(x) = x_1, \quad h(x) = x_2 - \frac{c_2}{x_1}\frac{a\tanh x_1 + x_1(z^0 - a)}{c_2 + kc_1x_1\tanh x_1} \tag{2.44}$$

the functional constraint set is given by

$$X = \{x \in \mathbb{R}^2 : f(x) \leq \frac{c_2}{c_1 v_0^2}, \; - f(x) \leq \frac{c_2}{c_1 v_1^2}, \; h(x) = 0\}. \tag{2.45}$$

† The same argument was given by Da Cunha and Polak [31].

Define the mappings $\bar{g}_i(\cdot): X \to \mathbb{R}$, $i = 1,2$, by

$$\bar{g}_1(x) = g_1(u^*(\cdot)) \quad \text{and} \quad \bar{g}_2(x) = g_2(u^*(\cdot)). \tag{2.46}$$

One then has a programming problem of the form: Obtain Pareto-optimal decisions $x^* = (x_1^*, x_2^*)$ for

$$\bar{g}_1(x) \text{ and } \bar{g}_2(x) \text{ subject to } x \in X. \tag{2.47}$$

In general the Pareto-optimal decisions will again be generated by the weighting parameter c. As before, if the problem has a unique solution for a fixed c, then Pareto-optimality follows. To simplify matters, choose $k = 1$, $a = 1$, $z^0 = 1$, and consider an equal weighting of the criteria with $c_1 = c_2 = 1$. The equality constraint may be substituted into the corresponding scalar combinations of the criteria to obtain a function of x_1 only, defined by

$$\bar{J}(x) = \bar{g}_1(x) + \bar{g}_2(x) = J(x_1) = \frac{1}{2x_1^3}\left[x_1 - \frac{\tanh x_1}{1 + x_1 \tanh x_1}\right]. \tag{2.48}$$

This is a strictly decreasing function of x with unique minimum obtained for the largest value, namely $x_1^* = \frac{1}{v_0}$.

Finally, the originally posed problem may be answered: The additional moment of inertia due to the device is to have its smallest value, or the device is to be placed as close as possible to the center of the disc, an intuitively expected answer.

3. NATURAL STRUCTURAL SHAPES

Optimal structural design is an old science, for certainly the circular shape of the wheel is "optimal" among all polygonal shapes which could be mounted on the axles of a cart; so much so that the word "wheel" has come to be synonymous with "circular shape." In ship design a certain amount of optimality has been achieved by the successive elimination of bad designs over thousands of years, the amazing part being that the basic shape has been around for such a long time. Currently one minimizes deflections, stresses, cost and a multitude of other criteria. In each case the designer must decide what is to be his criterion, his choice usually being based on experience or on a management decision. In this section a structural design criterion is proposed with resulting structures which exhibit a number of desirable characteristics. That is, the criterion is definitive for a class of structures with properties which should be attractive under a multitude of circumstances. In particular, it is assumed that a structure is to be designed with at least some broad purpose in mind and that the structure and its purpose can be defined in terms of a collection of parameters. The objective then is to devise a criterion which allows one to choose the parameters in an optimal manner. The proposed general theory is treated in detail in Stadler [28], which also includes several additional examples of the design of so-called natural structures. Here the treatment of the subject is purposely kept simple so as to make it accessible to a reader who has little or no background in structural mechanics. It is useful to begin with a fairly general notion of what comprises a structure.

3.1. Body and Structure. According to the evidence of our senses a body occupies a portion of space, it consists of matter, it has something which provides a resistance to moving it, it is cohesive, and finally, it has shape. Every construction of a mathematical model from a

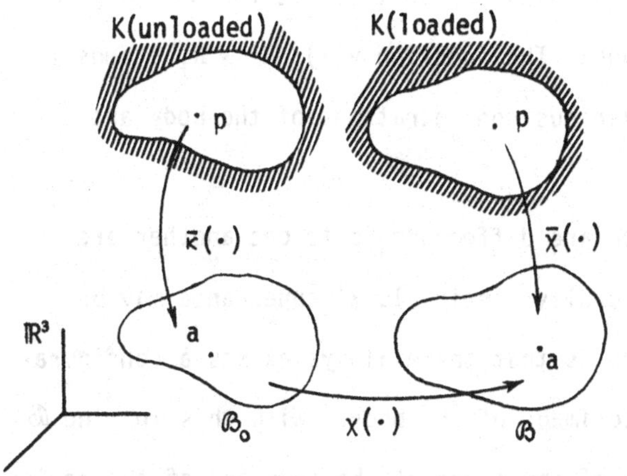

Fig. 2. The various configurations of the body K.

physical system involves the translation of the physical concepts into mathematical notions.† In the case of a body, rather than dealing with the physical body K one usually deals with it in terms of a subset $\mathcal{B}$ of $\mathbb{R}^3$; one accomplishes this change of thought basis by assuming that there exists a diffeomorphism $\bar{\chi}(\cdot)$: K $\to \mathcal{B}$, which maps the physical body K onto a subset $\mathcal{B}$ of $\mathbb{R}^3$. Such a mapping is also called a configuration of the body K. All properties of the body are then defined in terms of its image $\mathcal{B}$ in $\mathbb{R}^3$ under the mapping $\bar{\chi}(\cdot)$. In the treatment of static problems it is usual to take a particular configuration $\bar{\kappa}(\cdot)$ as a reference configuration with $\bar{\kappa}(K) = \mathcal{B}_0$. A deformation of the body then is a diffeomorphism $\chi(\cdot)$: $\mathcal{B}_0 \to \mathcal{B}$. In the remainder $\mathcal{B}_0$ will be called the reference (unloaded) shape of the body and $\mathcal{B}$ the deformed (as a result of "loading" the body) shape. The first property with which one usually endows the body K or any part $\bar{P}$ of it is its mass. This is done in terms of $\mathcal{P} \subset \mathcal{B}$ as

$$m(\mathcal{P}) = \int_{\mathcal{P}} \rho(x)dv(x),$$

† It is philosophically interesting to contemplate this as a mapping from one kind of thought to another kind of thought.

where $\rho(\cdot): \mathcal{B} \to \mathbb{R}_+$ is the mass density per unit volume for the shape $\mathcal{B}$[†] and where $\mathcal{P} = \overline{\chi}(\overline{P})$ is understood. The function $v(\cdot): \mathbb{R}^3 \to \mathbb{R}_+$ stands for Lebesgue measure on $\mathbb{R}^3$. The various configurations of the body are shown in Figure 2.

Mathematically, sets which are diffeomorphic to one another are indistinguishable locally; naturally, their global appearance may be different. The assumption here is that there always exists a configuration which provides an accurate image of the body. With this in mind $\mathcal{B}$ may be taken as representative of the shape of the body and of the region in space which the body occupies. The mass characterizes the body's resistance to motion. The remaining properties of the body are assumed to be completely characterized by a scalar function $\varepsilon(\cdot): \psi \times \mathcal{B} \to \mathbb{R}$ called the specific stored energy per unit volume. ψ is a parameter space; it is purposely left unspecified, for it may consist of function classes, subsets of some $\mathbb{R}^n$, etc. However, for any particular problem it is assumed to have been specified. In terms of $\varepsilon(\cdot)$, the total stored energy of K in the shape $\mathcal{B}$ is

$$\mathcal{U}(\mathcal{B}) = \int_{\mathcal{B}} \varepsilon(\psi, x)\, dv(x),$$

of course with a similar expression for each part $\overline{P}$ of K.

A little needs to be said about forces. Generally one uses "forces" to describe the unknown causes of certain effects. Two major classifications occur, contact forces and body forces. In particular, a distributed

† It is implicit in this statement that the density depends on the configuration.

contact force is representative of the effect that a part of the body has on an adjacent part $\overline{P}$ of the body. Let $\mathcal{J}$ be that part of the surface $\partial\mathcal{P}$ of $\mathcal{P}$ which is in contact with adjacent parts of the body (that is, $\overline{\chi}^{-1}(\partial\mathcal{P})$ is in contact), and let $n(\cdot):\mathcal{J} \to$ (the unit sphere in $\mathbb{R}^3$) be the unit outward normal on $\mathcal{J}$. Then the resultant contact force on the part $\mathcal{P}$ due to the adjacent parts is

$$f_c(\mathcal{P}) = \int_{\mathcal{J}} \hat{t}(x,n(x))d\mathcal{J},$$

where $\hat{t}(\cdot,n(\cdot)): \mathcal{J} \to \mathbb{R}^3$ is a suitable vector field and $t(x,n(x))$ is called the stress vector at $x \in \mathcal{J}$. Naturally, this expression also describes the distributed contact (contact over a non-zero part of the surface) force between two bodies; it does not describe a concentrated force.

A resultant external body force is similarly defined by

$$f_b(\mathcal{P}) = \int_{\mathcal{P}} b(x)dv(x),$$

where $b(\cdot): \mathcal{B} \to \mathbb{R}^3$ is the density per unit volume of the external body force. The word "external" refers to the fact that no contact is required for the force to act, the word "body" refers to the fact that it acts on each and every part of the body.

When a body is designed with a particular purpose in mind, then the result is called a structure.

Definition 3.1.1. Structure. Let the following be specified:

S1. The parameter space ψ, the specific stored energy $\varepsilon(\cdot)$, the reference shape $\mathcal{B}_0$ and the mass density $\rho(\cdot)$ in the shape $\mathcal{B}_0$;

S2. The body forces and the contact forces;

S3. A constitutive relation between the forces and the deformation;

S4. Restrictions on the kinds of possible deformations and possible geometric constraints;

S5. Equations which are definitive for the equilibrium (static, dynamic, thermal, etc.) of the body.

A particular choice for each definitive entity (parameters, equations, etc.) results in a collection S = {S1,...,S5}. S is called a structure. When some entities are allowed to vary within specified sets then this variation generates a subclass {S} of structures. □

In dealing with structures one is usually confronted with the questions: How much of a "load" can it support and what is the resulting deformation of the structure? The deformation can be treated in terms of the relative displacement of points $a \in \mathcal{B}_0$ -- the concept involves only geometry. To free the results from the influence of the particular shape of a structure, the concept of strain, an elongation per unit length, is introduced as a measure of the deformation.

In order to deal with distributed contact forces and in particular with the interaction forces between parts of the body, the concept of stress or a force per unit area is introduced. The particular form of the expression for the stress at a point in the body is derived from equilibrium considerations of a nearby part of the body.

Obviously, the interaction of force and deformation depends on the material, and one classifies ideal materials by constitutive relations, that is, equations which specify the stress as a function of the deformation.

Finally, one defines a boundary value problem in terms of the given forces, force or displacement boundary conditions and the requirement that the structure be in equilibrium. A detailed discussion of these concepts would lead too far; in summary then one needs to deal with:

(i) Kinematics. These involve the deformation $\chi(\cdot)$, its gradient $F(\cdot)$ and the concept of strain.

(ii) Kinetics. These involve the forces, moments, and their equipollence with a tensor field $T(\cdot): \mathcal{B} \to \mathbb{R}^3$, where $T(x)$ is called the stress tensor at $x \in \mathcal{B}$.

(iii) Constitutive Relations. The ultimate purpose of constitutive relations is to provide a link between the deformation and the forces and moments which are applied to the body. A simple such relation consists of the specification: For a suitable selection of the reference configuration, $F(a)F(a)^T = I$, the identity matrix, for every $a \in \mathcal{B}_0$. In other words, with a suitable choice of reference configuration the only possible deformation of the body is one which consists of a rotation about a point -- the body is a "rigid body."

(iv) Finally, a structure is in static equilibrium if the summation of the external forces (contact and body forces) and of the external moments on the body vanishes and if $T(\cdot)$ satisfies the equation

$$\operatorname{div} T(x) + \rho(x)b(x) = 0 \tag{3.1}$$

in the interior of $\mathcal{B}$ with specified conditions on $\partial\mathcal{B}$.

A large part of the preceding discussion is based on the fundamental work by Truesdell and Noll [32]. The remainder of the discussion is best formulated in terms of the particular example to be treated.

3.2. The Bar as a Natural Structure. The involved structure is an "axially loaded bar." Such structures appear, for example, as members in bridges or in building trusses. The definitive assumptions for such a structure are listed first, along with some corresponding illustrations.

Definition 3.2.1. Bar. A bar is a cylindrical three-dimensional solid body which is characterized by a set of equations and functions having the parameter of a specified curve as its only independent variable. □

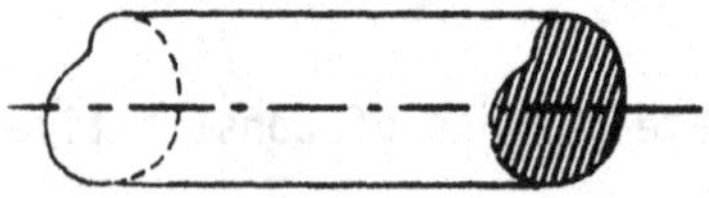

Fig. 3. A bar.

For a general theory of rods and bars, see Antman [33]; for simple axial extension members, see any basic text on the strength of materials.

Next, the definitive assumptions are stated.

A1. The curve in the Definition 3.2.1 is a straight line.

A2. Contact forces are applied only to the ends of the bar; there may be external body forces.

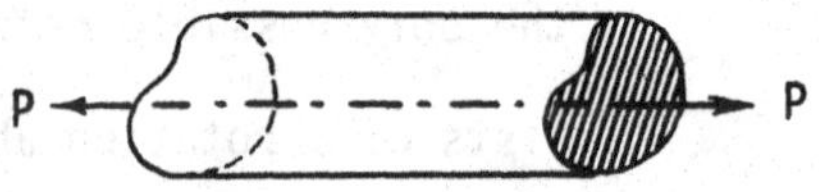

Fig. 4. A loaded bar.

A3. The deformation $\chi(\cdot)$ is sufficiently small so that the integral of any quantity $\phi(\cdot)$ over $\mathcal{B}$ may be replaced by the integral of the same quantity over $\mathcal{B}_0$. (For more detail, see Fox [29].)

A4. The total loading (body and contact forces) is such that it gives rise only to a uniform normal stress-distribution over

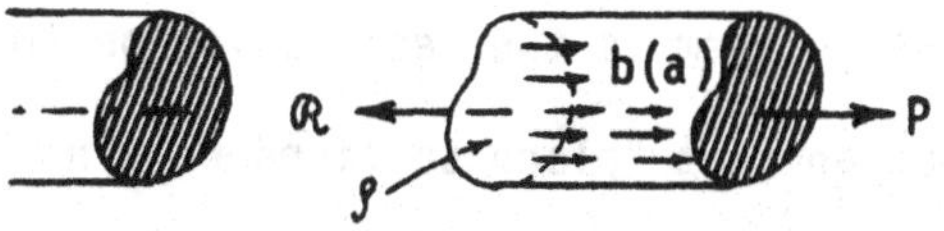

Fig. 5. The internal resultant.

each cross-section of the bar; that is, for example, with a unit outward normal n on 𝒮

$$\mathcal{R} = -\int_{\mathcal{S}} \tau(a)\,n\,d\mathcal{S} = -\tau(a)An, \tag{3.2}$$

if n is taken as positive to the right and if A is the total cross-sectional area. Thus, the stress tensor T(·) consists of a single component $\tau(\cdot): [0,L] \to \mathbb{R}$, for a bar of length L.

A5. The specific stored energy per unit volume is defined by

$$\sigma(a) = \frac{1}{2}E\left(\frac{dy}{da}(a)\right)^2, \tag{3.3}$$

where E is the elastic modulus and where $y(\cdot): [0,L] \to \mathbb{R}$, defined by

$$y(a) = \chi(a) - a, \tag{3.4}$$

is the displacement of a point $a \in \mathcal{B}_0$. In addition, the stress at a is related to σ(·) by

$$\tau(a) = \frac{d\sigma}{d(\frac{dy}{da})}(a) = E\frac{dy}{da}(a). \tag{3.5}$$

This particular constitutive relation is called Hooke's law. A larger class of materials with such a relation between the specific stored energy and the stress tensor T(·) is the class of hyperelastic materials, e.g., see Truesdell and Noll [32].

The previous assumptions are given for bars with constant cross-sections. They encompass what is usually called "engineering theory."

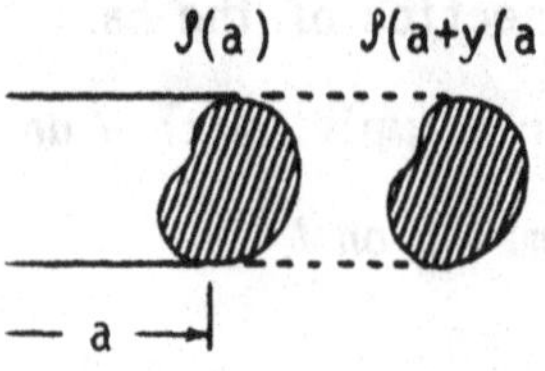

Fig. 6. Plane sections remain plane.

Implicit in them is the restriction on the deformation that "plane sections remain plane"; that is, a cross-section $\mathcal{J}(a)$ at a in the undeformed shape is displaced to a parallel section $\mathcal{J}(a+y(a))$ in the deformed shape. One additional assumption will be made.

A6. Variable cross-sections are permissible, that is, the cross-sectional area may be a function $A(\cdot)\colon [0,L] \to \mathbb{R}_+$.

The adequacy of the assumptions in representing the actual physical phenomenon will not be touched upon; the emphasis here is on the conceptual results rather than on physically quantitative ones. Some comparisons of the formulation of the same type of problem within the linear theory of elasticity may be found in Sokolnikoff [34]. Variable cross-sections will be of particular interest, since the objective in the example shall be to find among all possible variations in the area $A(\cdot)$ that one which is "optimal" in a yet to be defined sense. But consider first the following intuitive argument.

A bar is built in at the left end and has a constant cross-section, say $A(a) = A$ $\forall\ a \in [0,L]$. The bar is loaded only by a constant force P at the right end. For static equilibrium the summation of the forces on any part of the bar must be

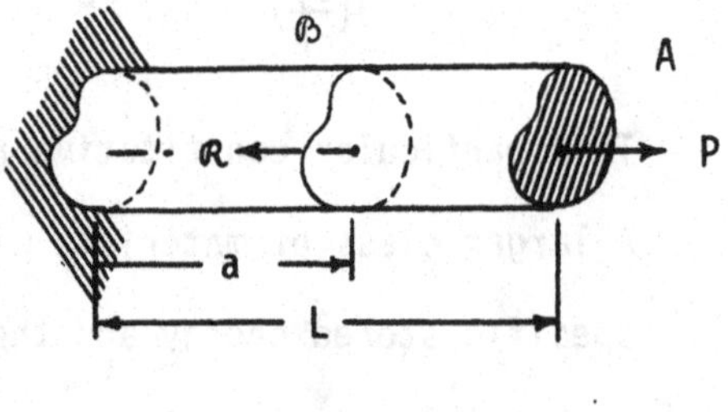

Fig. 7. A built in bar.

zero, that is,

$$\mathcal{R} = P \Rightarrow \tau(a) = \frac{P}{A} . \tag{3.6}$$

In view of A5 this is related to the displacement by

$$\frac{dy}{da}(a) = \frac{1}{E}\tau(a) = \frac{P}{EA} \Rightarrow y(a) = \int_0^a \frac{P}{EA} d\xi = \frac{Pa}{EA} , \tag{3.7}$$

since $y(0) = 0$. The total stored energy due to the displacement $y(\cdot)$ is

$$\mathcal{U}(\mathcal{B}) = \frac{1}{2}EA\int_0^L \left(\frac{P}{EA}\right)^2 d\xi = \frac{1}{2}\frac{P^2L}{AE} . \tag{3.8}$$

Now compare two bars $\mathcal{B}_1$ and $\mathcal{B}_2$, both built in at $a = 0$, of length L, and loaded by an axial load P with A and E not specified.

(1) Assume that $A_2 > A_1$ and that $E_1 = E_2 = E$. It follows that

$$y_2(a) < y_1(a) \quad \forall\ a \in [0,L] \tag{3.9}$$

and

$$\tau_2 = \frac{P}{A_2} < \tau_1 = \frac{P}{A_1} . \tag{3.10}$$

Furthermore,

$$\mathcal{U}(\mathcal{B}_2) < \mathcal{U}(\mathcal{B}_1). \tag{3.11}$$

(2) A similar conclusion follows with $E_1 < E_2$ and with $A_1 = A_2 = A$.

(3) The approach may be reversed, that is, one may start with the assumption

$$\mathcal{U}(\mathcal{B}_2) < \mathcal{U}(\mathcal{B}_1), \tag{3.12}$$

and note that it implies $(AE)_1 < (AE)_2$ and that hence, if $E_1 < E_2$, one again obtains

$$y_2(a) < y_1(a) \quad \forall\ a \in [0,L] \tag{3.13}$$

and

$$\tau_2 < \tau_1 \tag{3.14}$$

since $\frac{A_1}{A_2} < \frac{E_1}{E_2} < 1$.

Thus, if "best" design consisted of making the stresses and the displacements in the bar as small as possible, and if all the parameters except A and E were fixed with $\alpha \leqslant A \leqslant \beta$ and $\delta \leqslant E \leqslant \gamma$, then an optimal choice would obviously be $A = \beta$ and $E = \gamma$. Note that these decisions would also result in a minimum of the total stored energy. Coloquially then, one might call the bar $\mathcal{B}_2$ with stored energy $\mathcal{U}(\mathcal{B}_2)$ stronger than the bar $\mathcal{B}_1$.

All bars with built in left ends are a subclass of all structures. Any member of this subclass may be specified as a seven-tuple $B = \{E,\rho,y, A,L,b,P\}$, with the subclass denoted by $\{B\}$. Clearly, some of the parameters in $\{B\}$ may be functions of the independent variable a, or possibly of each other if some additional constraints are imposed. Naturally, any of the parameters could be specified as fixed in a given problem, with the remaining ones used as "controls" to find a "best" suited one in a given subclass.

<u>Definition 3.2.2</u>. Strength of a structure. Let $\{S\}$ be a given subclass of structures all of whose members are in static equilibrium. Let $S_1, S_2 \in \{S\}$. Then the structure S_1 is stronger than the structure S_2 iff the corresponding stored energies and reference shapes are related by

$\mathcal{U}(\mathcal{B}_{01}) < \mathcal{U}(\mathcal{B}_{02})$. □

Generally, a "strongest" structure cannot be obtained unless the problem is suitably restricted; e.g., up to a point it may be possible to make a structure stronger by assigning more bulk. However, the addition of bulk usually changes only the size of the structure, not its outline; i.e., one may have a ten-foot diameter sphere or a twenty-foot diameter sphere. Thus, one might expect mass to play a crucial role in the optimal design (an optimal choice of the variable parameters) of a structure.

In the following, only structures belonging to a given subclass are to be compared. It is assumed that in the final stage of defining a structure the statements S1 - S5 have been reduced to a collection of n parameters. Some of these will be given a priori, some fixed by necessity or choice, say a total of k of them. Let the remaining n-k parameters be designated as controls $u_i(\cdot): Q_i \to \mathbb{R}$, $i = 1,\ldots,n-k$, where the Q_i are suitably chosen domains. An optimal structure is then defined in terms of these controls for the criteria which are given next. For simplicity the definition given here is restricted to "one-dimensional" structures, i.e., the controls are functions of a common single independent variable, with only small deformations to be considered. Let

$$g_1(u(\cdot)) = \int_{\mathcal{B}_0} \rho(a)dv(a) \quad \text{and} \quad g_2(u(\cdot)) = \int_{\mathcal{B}_0} \varepsilon(a)dv(a), \tag{3.15}$$

where the remaining variables in the specific stored energy have been suppressed.† It is assumed that the class $\mathcal{F}$ of admissible controls is non-empty.

<u>Definition 3.2.3</u>. Natural structure. Assume that {S} is a subclass of structures all of whose members are in static equilibrium. Let

† Normally these integrations should be carried out over $\mathcal{B}$; $\mathcal{B}_0$ is used here as a consequence of Assumption A3. For a discussion of the meaning of "small deformations" see reference 29.

$u(\cdot) \in \mathfrak{F}$ be the collective expression for those parameters which are to be used as controls. Then $S^* \in \{S\}$ is a natural structure iff $u^*(\cdot) \in \mathfrak{F}$ is a Pareto-optimal control for $g_1(u(\cdot))$ and $g_2(u(\cdot))$ with both referred to $\{S\}$. The corresponding shape $\mathcal{B}_0^*$ is a natural shape. □

<u>3.3. The Stalactite.</u> Consider an axially loaded bar which is built in at the top end and loaded by its own weight and a constant load Q at the free end. The length L, the density ρ and the modulus E are assumed to be given constants. The loading on part I due to the weight of part II and Q is

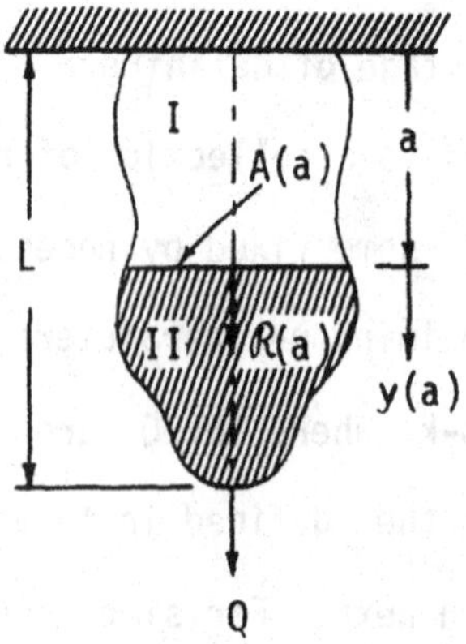

Fig. 8. A bar loaded by its own weight.

$$\mathcal{R}(a) = Q + \rho g \int_a^L A(\xi)d\xi \,, \tag{3.16}$$

that is, the cross-sectional area is taken to be a function $A(\cdot): [0,L] \to \mathbb{R}$. Quite generally, in view of the fact that the criteria $g_1(u(\cdot))$ and $g_2(u(\cdot))$ are non-commensurate, it is convenient to deal with non-dimensional quantities throughout. Towards this purpose introduce

$$t = \frac{a}{L}, \quad x_1(t) = \frac{y(tL)}{L}, \quad \omega = \frac{Q}{P}, \quad u(t) = \frac{A(tL)}{L^2},$$
$$x_2(t) = \omega + \frac{L^3\rho g}{P}\int_t^1 u(\xi)d\xi \,, \tag{3.17}$$

where P is some arbitrary constant force.

One then has a standard problem of Type II with Pareto-optimality as the optimality concept. In the present context the problem may be stated as: Obtain the natural shape $\mathcal{B}_0^*$ for the system characterized by

$$g_1(u(\cdot)) = \int_0^1 u(\xi)d\xi \quad \text{and} \quad g_2(u(\cdot)) = \int_0^1 \frac{x_2^2(\xi)}{u(\xi)}d\xi \tag{3.18}$$

subject to

$$\begin{aligned} \dot{x}_1 &= k_1\frac{x_2}{u}, \quad x_1(0) = 0, \quad k_1 = \frac{P}{L^2E}, \\ \dot{x}_2 &= -k_2u, \quad x_2(0) = \omega, \quad k_2 = \frac{L^3\rho g}{P}. \end{aligned} \tag{3.19}$$

The H-function for this problem is

$$H(\lambda,x,u) = -c_1u - c_2\frac{x_2^2}{u} + \lambda_1k_1\frac{x_2}{u} - \lambda_2k_2u, \tag{3.20}$$

with adjoint equations

$$\begin{aligned} \dot{\lambda}_1 &= 0, \\ \dot{\lambda}_2 &= 2c_2\frac{x_2}{u} - \lambda_1\frac{k_1}{u}. \end{aligned} \tag{3.21}$$

The transversality conditions yield $\lambda_2(0) = 0$, and $\lambda_1(1) = 0$, so that $\lambda_1(t) = 0$. From $\frac{\partial H}{\partial u} = 0$ it follows that

$$u^*(t) = \frac{\sqrt{c_2}x_2(t)}{\sqrt{\lambda_2(t)k_2 + c_1}}. \tag{3.22}$$

The use of

$$\dot{\lambda}_2(t) = 2\sqrt{c_2}\sqrt{c_1 + k_2\lambda_2(t)}, \tag{3.23}$$

or, upon integration thereof,

$$\sqrt{k_2\lambda_2(t) + c_1} = k_2\sqrt{c_2}t + \sqrt{c_1}, \tag{3.24}$$

results in

$$u^*(t) = \frac{\sqrt{c_2}x_2(t)}{k_2\sqrt{c_2}t + \sqrt{c_1}}. \tag{3.25}$$

The substitution of $u^*(t)$ in the state equations (3.19) yields

$$x_2^*(t) = \omega \frac{\sqrt{c_2}k_2 + \sqrt{c_1}}{\sqrt{c_2}k_2 t + \sqrt{c_1}} \tag{3.26}$$

and

$$x_1^*(t) = \frac{k_1}{2k_2}\left[\left(k_2 t + \sqrt{\frac{c_1}{c_2}}\right)^2 - \frac{c_1}{c_2}\right] \tag{3.27}$$

for the displacement, corresponding to an area distribution

$$u^*(t) = \omega \frac{c_2 k_2 + \sqrt{c_1 c_2}}{(\sqrt{c_2}k_2 t + \sqrt{c_1})^2} . \tag{3.28}$$

If a circular cross-section with non-dimensional radius $e(t) = r(tL)/L$ is assumed, then one obtains

$$e^*(t) = \frac{1}{\sqrt{\pi}} \frac{\sqrt{\omega(k_2 c_2 + \sqrt{c_1 c_2})}}{k_2\sqrt{c_2}t + \sqrt{c_1}} . \tag{3.29}$$

Remark 3.1. Appropriately one should now check sufficient conditions to establish Pareto-optimality of the control (3.28). It seems more appropriate here to provide a comparison with another "optimal" design, since industrial applications will generally be more concerned with simply obtaining a "better" design rather than the elusive assured "best".

Fig. 9. Stalactites.

The functional form of the radius makes clear the choice of the title "stalactite".

3.4. A Comparison with an Optimal Design. Another concept of an optimal design is a structure with a constant stress-distribution throughout the structure, the implication being that one makes full use of all parts of the structure in carrying the load. This design has the disadvantage that the yields simultaneously at every point, but it usually provides a minimum weight design.

For comparison, consider then a bar subjected to the same loading as the stalactite, but let the design criterion be the attainment of a constant, nondimensional stress $\bar{\tau}$ at each section; denote all quantities concerned with the constant stress design by affixing an overbar to the symbols. Then

$$\bar{\tau}(t) = \frac{\bar{x}_2(t)}{\bar{u}(t)} = \frac{1}{\bar{u}(t)}[\omega + k_2\int_t^1 \bar{u}(\xi)d\xi] = \bar{\tau} = \text{const.} \tag{3.30}$$

Consequently,

$$\frac{d\bar{u}}{dt}(t) + \frac{k_2}{\bar{\tau}}\bar{u}(t) = 0 \tag{3.31}$$

and

$$\bar{u}(t) = \frac{\omega}{\bar{\tau}}\exp\frac{k_2}{\bar{\tau}}(1-t). \tag{3.32}$$

This results in a loading

$$\bar{x}_2(t) = \omega \exp\frac{k_2}{\bar{\tau}}(1-t) \tag{3.33}$$

and a deformation

$$\bar{x}_1(t) = k_1\bar{\tau}t. \tag{3.34}$$

A comparison of the two designs will first be made for a maximum allowable stress. Assume that the maximum allowable stress for the suspended bar is $\hat{\tau}$. The maximum nondimensional stress for the natural structure occurs at the free end; hence,

$$\tau^*(1) = \frac{x_2^*(1)}{u^*(1)} = \frac{1}{c_2}(k_2c_2 + \sqrt{c_1c_2}) = \hat{\tau} \tag{3.35}$$

and consequently,

$$g_1(u^*(\cdot)) = \frac{\omega}{\hat{\tau} - k_2}, \tag{3.36}$$

$$g_2(u^*(\cdot)) = \omega\hat{\tau}, \tag{3.37}$$

$$x_1^*(1) = \frac{1}{2}k_1(2\hat{\tau} - k_2). \tag{3.38}$$

For the constant stress structure one has

$$g_1(\bar{u}(\cdot)) = \frac{\omega}{k_2}(\exp\frac{k_2}{\hat{\tau}} - 1), \tag{3.39}$$

$$g_2(\bar{u}(\cdot)) = \frac{\omega\hat{\tau}^2}{k_2}(\exp\frac{k_2}{\hat{\tau}} - 1), \tag{3.40}$$

$$\bar{x}_1(1) = k_1\hat{\tau}. \tag{3.41}$$

As one might have expected

$$g_1(\bar{u}(\cdot)) < g_1(u^*(\cdot)). \tag{3.42}$$

There is, however, a price to pay in terms of the amount of resulting deformation. This is evidenced in that

$$g_2(u^*(\cdot)) < g_2(\bar{u}(\cdot)), \tag{3.43}$$

i.e., the natural structure is stronger. Furthermore,

$$x_1^*(1) < \bar{x}_1(1), \tag{3.44}$$

so that for the same maximum allowable stress, the maximum extension of the natural structure is less than that of the constant stress structure.

As a final comparison, let the maximum allowable extension of the bar be specified as δ. For the natural structure this requires

$$x_1^*(1) = \frac{k_1}{2c_2}(k_2c_2 + 2\sqrt{c_1c_2}) = \delta \tag{3.45}$$

and a bulk

$$g_1(u^*(\cdot)) = \omega\sqrt{\frac{c_2}{c_1}} = \frac{2k_1\omega}{2\delta - k_1k_2} \tag{3.46}$$

to sustain it. For the constant stress structure

$$\bar{x}_1(1) = k_1\bar{\tau} = \delta \tag{3.47}$$

so that the corresponding necessary bulk is

$$g_1(\bar{u}(\cdot)) = \frac{\omega}{k_2}(\exp\frac{k_1k_2}{\delta} - 1). \tag{3.48}$$

Thus, one always has

$$g_1(u^*(\cdot)) < g_1(\bar{u}(\cdot)), \tag{3.49}$$

as a matter of fact, the amount of material saved by using a natural bar increases with a decrease in the allowable extension δ.

3.5. Conclusion. Some examples of the application of the concept of Pareto-optimality in Mechanics have been given. The objective in the first example was to minimize the disturbance to an optimally controlled system due to the introduction of measuring devices. It seems that such a consideration would enter into the design of any feedback system where the current state is to be measured with a consequent disturbance to the system. Naturally, the concept has applications in other engineering and nonengineering fields; in particular, it should be of further interest in electrical engineering where small disturbances can cause considerable alterations of the system.

The second application concerned optimal structural design. There are many problems for which the optimality criterion is predetermined by the result one wishes to attain; that is, the objective might be to minimize the weight, a particular deflection, or the natural frequency of the structure, and so on. This type of problem was not considered here or in Stadler [28], rather it was the intent to present a whole class of structures whose properties are desirable under any circumstances. Another example which gave impetus to the name "natural structure" was that of a branch, which is the result in the calculation of the natural shape of a cantilever beam loaded only by its own weight. This example and more are given in the same reference. It seems desirable to carry on additional investigations concerning the properties of such natural structures. The fundamental concepts of mass and stored energy appear directly or in some equivalent form in many other fields; thus, the same sort of "structure" would appear in any of these fields with the corresponding, possibly desirable, properties. Hence, a further extension of the theory in these directions would be appropriate.

ACKNOWLEDGEMENTS

Basically, gratitude is a state of mind expressed continuously in our actions. Sometimes there is a need to formally express it. I am grateful to G. Leitmann for all that I have learned from him throughout many discussions and for his critical reading of much of my work. I appreciate the patience of my friends Majdenin, Hajimi, Wen-Hon, Santiago, Hadi, Sol and Norbert for listening to my discourses and for making helpful suggestions in writing this paper, and the kindness of Professors A. Marzollo and G. Leitmann in providing the opportunity to present this work at CISM.

Naturally, I am also grateful for the financial support of CISM for the presentation of these lectures at CISM, and for that of the Office of Naval Research in writing some of the material. Thanks also to Ms. Donahue for typing the first draft and to Spatz for the final manuscript.

REFERENCES

1. Yu, P.L. and Leitmann, G., Compromise solutions, domination structures and Salukvadze's solution, *J. of Optimization Theory and Applications*, 13, 362, 1973.

2. Stadler, W., Preference optimality, to be published.

3. Stadler, W., Sufficient conditions for preference optimality, to appear in *J. of Optimization Theory and Applications*, 1975.

4. Lee, E.B. and Markus, L., *Foundations of Optimal Control Theory*, John Wiley and Sons, New York, 1967.

5. Debreu, G., *Theory of Value*, John Wiley and Sons, New York, 1959.

6. Samuelson, P.A., *Economics*, ninth ed., McGraw-Hill, New York, 1969.

7. Mangasarian, O.L., *Nonlinear Programming*, McGraw-Hill, New York, 1969.

8. Fishburn, P.C., *Utility Theory for Decision Making*, John Wiley and Sons, 1970.

9. Debreu, G., Smooth preferences, *Econometrica*, 40, 603, 1972.

10. Matsushima, Y., *Differential Manifolds*, translated from the Japanese by Kobayashi, E.T., Marcel Dekker, Inc., New York, 1972.

11. Spivak, M., *Differential Geometry*, I and II, Publish or Perish, Inc., 6 Beacon St., Boston, Mass. 02108 (USA), 1970.

12. Debreu, G., Representation of a preference ordering by a numerical function, in *Decision Processes*, Thrall, R.M., Coombs, C.H. and Davis, R.L., Eds., John Wiley and Sons, New York, 1954.

13. Fishburn, P.C., Lexicographic orders, utilities and decision rules: a survey, *Management Science*, 20, 1442, 1974.

14. Eilenberg, S., Ordered topological spaces, *Amer. J. of Math.*, 63, 39, 1941.

15. Rader, J.T., The existence of a utility function to represent preferences, *The Review of Economic Studies*, 30, 229, 1963.

16. Debreu, G., Continuity properties of Paretian utility, *Int'l Economic Review*, 5, 285, 1964.

17. de Finetti, B., Sulle stratifiazioni convesse, *Annali di Matematica Pura ed Applicata*, 4, 173, 1949.

18. Fenchel, W., *Convex Cones, Sets and Functions*, mimeographed notes, Department of Mathematics, Princeton University, 1953.

19. Fenchel, W., Uber konvexe Funktionen mit vorgeschriebenen Niveau mannigfaltigkeiten, *Mathematische Zeitschrift*, 63, 496, 1956.

20. Long, R.S., Newton-Raphson operator; problems with undetermined end points, *J. of the Amer. Inst. of Aeronautics and Astronautics*, 3, 1351, 1965.

21. Leitmann, G., *Einführung in die Theorie optimaler Steuerung und der Differentialspiele*, translated from the English by Stadler, W., R. Oldenbourg Verlag, Munich, 1974.

22. Willard, S., *General Topology*, Addison-Wesley, Reading, Mass., 1970.

23. Yu, P.L., Cone convexity, cone extreme points, and nondominated solutions in decision problems with multiobjectives, *J. of Optimization Theory and Applications*, 14, 319, 1974.

24. Yu, P.L. and Leitmann, G., Nondominated decisions and cone convexity in dynamic multicriteria decision problems, *J. of Optimization Theory and Applications*, 14, 573, 1974.

25. Aumann, R.J., Values of markets with a continuum of traders, Technical Report no. 121, *The Economics Series*, Institute for Mathematical Studies in the Social Sciences, Stanford University, Stanford, 1974.

26. Salama, A.I.A. and Gourishankar, V., Optimal control of systems with a single control and several cost functionals, *Int'l J. of Control*, 14, 705, 1971.

27. Leitmann, G. and Stadler, W., Cooperative games for the experimentalist, *Nonlinear Vibration Problems* (Zagadnienia Drgań Nieliniowych), Polish Academy of Sciences, 15, 273, 1974.

28. Stadler, W., Natural structures, in preparation.

29. Fox, E.A., *Mechanics*, Harper and Row, New York, 1967.

30. Athans, M. and Falb, P.L., *Optimal Control*, McGraw-Hill, New York, 1966

31. Da Cunha, N.O. and Polak, E., Constrained minimization under vector-valued criteria in finite dimensional spaces, *J. of Mathematical Analysis and Applications*, 19, 103, 1967.

32. Truesdell, C. and Noll, W., *The Nonlinear Field Theories of Mechanics*, Handbuch der Physik, III/3, Springer Verlag, 1965.

33. Antmann, S.S., *The Theory of Rods*, Handbuch der Physik, VIa/2, Springer Verlag, 1972.

34. Sokolnikoff, I.S., *Mathematical Theory of Elasticity*, McGraw-Hill, New York, 1956.

DOMINATION STRUCTURES AND NONDOMINATED SOLUTIONS

P. L. Yu
Graduate School of Business
The University of Texas
Austin, Texas 78712

1. INTRODUCTION TO DOMINATION STRUCTURES

It is a well-known fact that decision makers are often faced with making decisions involving more than one criterion. Although every decision maker eventually makes his decision based on his intuition or judgement, it does not mean that he cannot benefit from a systematic analysis of his decision problem. In order to aid him in reaching a "good" decision, a number of concepts have been introduced, such as satisfaction, efficiency, utility construction, compromise solution, chance constraints, goal programming, and generalizations such as domination structures and nondominated solutions.

In this talk we shall focus on domination structures and nondominated solutions which were first introduced in [36]. We shall discuss the relationships between these concepts and other solution concepts. Some main characteristics of nondominated solutions will be discussed. Computational methods to obtain solutions as well as applications of the domination structure concept will be described.

In order to facilitate our discussion, we consider the following stock investment problem (from now on we shall use the letters SIP to refer to this problem): An investor wants to invest \$M in n stocks for a year. Assume that the expected return and the variance of the return for each stock are the only information available to him. How should he make his decision? This is a typical example of a multicriteria decision problem. Observe that this problem involves two basic elements: a set of choices (all possible allocations of \$M among n stocks) and a set of criteria (the expected return and the variance of the return) for each choice. In fact every multicriteria decision problem has these two elements. A more detailed and extensive discussion of SIP can be found in [38]. For the convenience of later discussion, we shall use the following notation.

Definition 1.1. We shall use X to denote the set of all feasible choices of our decision problem and call it the decision space. An element of X will be denoted by x. The criteria function (defined for all $x \in X$) will be denoted by $f(x) = (f_1(x),\ldots,f_\ell(x))$. The set $Y = \{f(x) | x \in X\}$ will be called the criteria space. An element of Y will be denoted by y. We shall assume that Y contains at least two distinct points; otherwise, our decision problem is not interesting.

Remark 1.1. It is usually very difficult to visualize the precise shape of Y. However, with certain conditions on X and f(x), we can study the cone convexity of Y. (See Section 2). We shall study decision making rationale based on Y, although the actual decision must be a point of X.

Example 1.1. In SIP, we can let $(m_1,\ldots,m_n)$ with $m_i \geqq 0$ and $\sum_i m_i = M$ denote an investment policy. Let $x_i = m_i/M$. We see that

$$X = \{(x_1,\ldots,x_n) \mid x_i \geqq 0, \sum_i x_i = 1\}. \tag{1.1}$$

Let $x = (x_1,\ldots,x_n)$ and

$$f_1(x) = \text{the variance}^{\dagger} \text{ of the return from policy } x \tag{1.2}$$

$$f_2(x) = \text{the expected return from } x \tag{1.3}$$

Then the criteria space of the SIP is given by

$$Y = \{(f_1(x), f_2(x)) \mid x \in X\}.$$

Now, given X and Y, how do we make a good decision? There are a number of well-known solution concepts for this problem discussed in the literature. Although each solution concept has its merits and shortcomings, none has been accepted universally (for a survey see [21]). We can classify the existing solution concepts as follows.

(i) One dimensional comparison: In this class of solution concepts, one first constructs a real-valued function u(y) defined on Y. Then an extreme value u(y) (maximum or minimum) over Y is located for the decision. In the case of maximizing u(y), u(y) is referred to as a utility function, efficiency index or preference ordering. Much research has been devoted to

†Note that as a consequence, as $f_1(x)$ increases, the variance of the return from choice x decreases.

the construction of such a utility function. [See [7]]. In the case of minimizing u(y), u(y) may be interpreted as a regret function or a function representing the distance from the utopia or ideal point. Compromise solutions and goal programming belong to this class of solution concepts. [See 4, 35]. The concepts of minimax and maximin also belong to the category of one dimensional comparison solution concepts.

(ii) Multiple dimension comparison: A solution x^0 is efficient or Pareto optimal if there is no other feasible solution x so that $f_i(x) \geq f_i(x^0)$ for each $i = 1,\ldots,\ell$ and $f(x) \neq f(x^0)$ where $f = (f_1,\ldots,f_\ell)$. Let $(F_1(y),\ldots,F_p(y))$ be a set of real-valued functions defined on the criteria space Y. Then x^0 is functionally efficient with respect to $(F_1,\ldots,F_p)$ if there is no other feasible x such that $F_i(f(x)) \geq F_i(f(x^0))$ for each $i = 1,\ldots,p$, and $F(f(x)) \neq F(f(x^0))$ where $F = (F_1,\ldots,F_p)$ [See [4]]. This concept of efficiency or Pareto optimality has been used quite extensively in the social sciences. These solutions are also called noninferior points or admissible strategies in other contexts. [See [6]].

(iii) Satisficing Models: In this approach the decision maker first establishes either

1) a minimal satisfaction level for each criteria or

2) an upper "goal achievement" level for each criterion.

In the first case a decision which does not exceed the minimal level for each criteria is unacceptable and will not be considered as a possibility for the final decision. In the second case any decision which exceeds the established upper level for each criteria is an acceptable final decision.

(iv) Ordering and Ranking: Instead of defining a real-valued function over Y as in (i), we define a binary relation which may or may not be a partial ordering over Y. We then find the resulting maximum or minimum elements over Y whenever they are well defined. Note that (ii) may be regarded as a special case of (iv). See [26,31].

(v) One at a time and iterative procedure: We first order the criteria according to their importance to the decision-maker such that if $i < j$ then f_i is more important than f_j. We then maximize f_1 over X. If this maximal solution is now unique, it will be used for the decision. Otherwise we maximize f_2 over those points which maximize f_1 over X. If this maximal solution is now unique, it will be used for the decision. Otherwise we maximize f_3 over those points which maximize both f_1 and f_2. The procedure is repeated until a unique solution is obtained or all the criteria have been considered. This solution approach is called a lexicographic ordering. Another iterative method has been used in the concept of satisficing. The idea is to start with a low satisfaction level for each criterion. If there are more than one solution the satisfaction level will be raised iteratively until a final decision can be obtained. Note that "gradient search" [13] can be put in this category too.

Clearly, the above concepts can be mixed and used simultaneously. For instance, combining (i) and (iii), we have a mathematical programming problem where the objective to be maximized is specified by (i) and the constraints are given by (iii). In the case where the upper goal achievement level of each criterion are known and there are no available solutions which exceed each upper level, we can construct a regret or distance function. This function will specify the distance from each feasible y to the set of points which satisfy the upper level of each criteria. This will result in a combination of (iii) and (i) again. Other combinations such as (i) and (ii), (i) and (iv), (i) and (v), (ii) and (vii), (ii) and (v), etc., are certainly possible. We shall leave these to the reader's imagination.

We do not intend to make a survey of the literature. However we hope that the above summary will help our discussion. We shall return to these solution concepts after we introduce the following concepts of domination structures and nondominated solutions.

Given two outcomes, y^1 and y^2, in Y, we can write $y^2 = y^1 + d$. If y^1 is preferred by the decision maker to y^2, written $y^1 \succ y^2$, we can think of this preference as occuring because of the factor d.

Definition 1.2. A nonzero vector d is a domination factor for $y \in Y$ if $y \succ y + \lambda d$ for all $\lambda > 0$. The set of all domination factors for y together with the zero vector will be denoted by D(y). The family $\{D(y) | y \in Y\}$, denoted simply by $D(\cdot)$, is called the domination structure of our decision problem.

Remark 1.2. By the definition, given a domination factor, d, for y, then any positive multiple of d is also a domination factor for y. It follows that, given $y^1 \succ y^2$, it is not necessarily true that $d = y^2 - y^1$ is a domination factor for y^1. Intuitively, one may regard a domination factor as a "bad" factor (thus, any positive multiple of it is also bad).

Definition 1.3. Given Y, $D(\cdot)$ and y^1, y^2 of Y, by y^1 is dominated by

y^2 we mean $y^1 \varepsilon y^2 + D(y^2) = \{y^2 + d | d \varepsilon D(y^2)\}$. A point $y^0 \varepsilon Y$ is a <u>nondominated solution</u> (or <u>nondominated outcome</u>) if there is no $y^1 \varepsilon Y$, $y^1 \neq y^0$, such that $y^0 \varepsilon y^1 + D(y^1)$. Thus y^0 is nondominated if and only if it is not dominated by any other outcome. Likewise, in the decision space S, a point $x^0 \varepsilon X$ is a nondominated solution (or <u>nondominated decision</u>) if there is no $x^1 \varepsilon X$ so that $f(x^0) \neq f(x^1)$ and $f(x^0) \varepsilon f(x^1) + D(f(x^1))$. The set of all nondominated solutions in the decision and criteria space will be denoted by N_X $(D(\cdot))$ and N_Y $(D(\cdot))$ respectively (or N_X and N_Y respectively when $D(\cdot)$ is unambiguously specified).

One important class of domination structures is characterized by the condition that $D(y) = \Lambda$, Λ is a convex cone for all $y \varepsilon Y$. In this case, we shall call Λ the <u>domination cone</u>. Because of its geometric significance, we have

<u>Definition 1.4</u>. A nondominated solution with respect to a domination cone Λ is called a <u>Λ-extreme point</u>. The set of all Λ-extreme ("cone-extreme") points is denoted by Ext $[Y|\Lambda]$.

<u>Example 1.2</u>. The additive weight method is an important approach in one dimensional comparison category (i). One first finds a <u>weight vector</u> (or <u>weight ratio</u>) $\lambda = (\lambda_1,\ldots,\lambda_\ell)$ and then maximizes $\lambda\cdot f = \sum_{i=1}^{\ell} \lambda_i f_i(x)$ over X. We see that this method implicitly uses a constant domination cone $\Lambda = \{d \varepsilon R^\ell | \lambda\cdot d < 0\}$ for each $y \varepsilon Y$. Note that Λ is a half space. Thus in order to use this method, a very strong assumption on the preference structure has to be imposed. Once Λ is revealed, it is not difficult to select the solutions which maximize $\lambda\cdot f$, or equivalently the nondominated solutions with respect to the domination cone Λ.

Example 1.3. To specify the weight ratio precisely for the additive weight method usually is a very difficult task. A more practical and realistic way to proceed is to require the weight ratio or weight vector to lie in a specified convex set. For instance, for SIP we may require that (λ_1,λ_2) satisfy $\frac{1}{2} < \lambda_2/\lambda_1 < 2$. This is an important revelation of preferences. As will be seen in Remark 2.1, if we set

$$\Lambda = \{(d_1, d_2) \left| \begin{array}{l} d_1 + 2d_2 \leq 0 \\ 2d_1 + d_2 \leq 0 \end{array} \right.\}$$

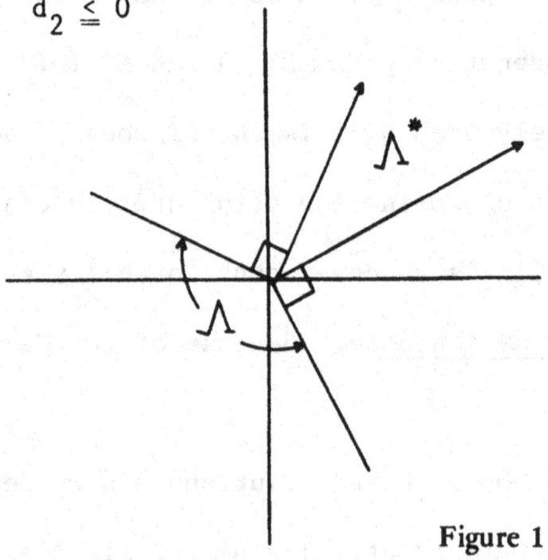

Figure 1

(see Figure 1), then all optimal solutions resulting from the additive weight method with the bounds $\frac{1}{2} < \lambda_2/\lambda_1 < 2$ will be contained in $\text{Ext}[Y|\Lambda]$. Thus in terms of the final decision which is generated, specifying the bounds of the weight vectors as above is "equivalent" to specifying a constant domination cone Λ. (See Remark 2.1 for further discussion.)

Example 1.4. Suppose that we specify a pseudo-concave utility function $U(y)$ with $\nabla U(y) \neq 0$ over Y. Then, implicitly, we have assumed a domination structure with $D(y)$ being bounded by: $\{d \in R^{\ell} | \nabla U(y)\cdot d < 0\} \subset D(y) \subset \{d \in R^{\ell} | \nabla U(y)\cdot d \leq 0\}$. Note that if y^0 maximizes $U(y)$ over Y, then y^0 is nondominated with respect to D(·). Also observe that each D(y) is at least a half space and in contrast to Example 1.2, D(y) varies with y. Note that the requirement that each D(·) is at least a half space is implicitly assumed in the revealed preference theory. (For instance see [15]).

Example 1.5. Let $\Lambda^{\leqq} = \{d \in R^{\ell} | d \leqq 0\}$. We see that y is Pareto optimal or efficient if and only if y is a $\Lambda^{\leqq}$-extreme points. That is, in the concept of Pareto optimality one uses a constant domination cone $\Lambda^{\leqq}$. Observe that $\Lambda^{\leqq}$ is only $1/2^{\ell}$ of the entire space. When $\ell = 6$ for instance, $\Lambda^{\leqq}$ is only 1/64 of R^6. (Of course, in this case, $\Lambda^{\leqq}$ is much smaller than a half space.)

Remark 1.3. In one dimensional comparison, suppose that the set Y* of optimal solutions of U(y) over Y is a convex set. We can easily construct a domination structure so that each optimal solution is a nondominated solution with respect to this domination structure. When U(y) has some special concavity properties which are usually assumed (see Example 1.4), the domination structure can be constructed in such a way that each D(y) contains at least a half space for each $y \in Y$ except in the interior Y*. This half space requirement on D(y) for each y is usually very difficult to fulfill and makes the solution concept difficult to be accepted or applied.

Remark 1.4. As shown in Example 1.5, the Pareto optimal or efficient solution concept is associated with a constant domination cone $\Lambda^{\leqq}$ which is only $1/2^{\ell}$ of the entire space, R^{ℓ}. In this approach, we do not take advantage of any additional partial information concerning the decision maker's preferences. For instance, it is not unusual that the decision maker may be able to specify bounds on the ratio of the weight vectors (see Example 1.3). This information will reveal a constant domination cone which usually is much larger than $\Lambda^{\leqq}$. In our later discussion we shall show how we can regard domination cones as "information sets" which characterize the decision maker's preferences; we also show how to select domination cones for the decision making when preferences are partially known.

Remark 1.5. Given a domination structure $D(\cdot)$ on Y, for $y^1, y^2 \in Y$ we can define $y^1 \succ y^2$ if and only if $y^2 \in y^1 + D(y^1)$. If in all $y \in Y$, $D(y) = \Lambda$, a constant domination cone containing no subspace other than $\{0\}$, then "$\succ$" is a partial ordering on Y. However, in general, the relation "$\succ$" is not necessarily transitive. (Also see (iv) ordering and ranking.)

Remark 1.6. As it will be shown later, nondominated solutions are closely related to the additive weight method (one dimensional comparison) and satisficing models. We shall show that each nondominated solution must be a unique maximum solution of an objective function which is a linear combination of the criteria subject to a set of constraints which are also linear combinations of the criteria. The objective and the constraints are in fact interchangeable.

The following are some basic properties enjoyed by nondominated solutions.

Theorem 1.1 (Lemma 4.1. of [36])

(i) $\text{Ext}\,[Y|\Lambda] = \begin{cases} Y & \text{if } \Lambda = \{0\} \\ \phi & \text{if } \Lambda = R^{\ell} \end{cases}$

(ii) If $\Lambda_1 \subset \Lambda_2$ then $\text{Ext}\,[Y|\Lambda_2] \subset \text{Ext}\,[Y|\Lambda_1]$

(iii) $\text{Ext}\,[Y+\Lambda|\Lambda] \subset \text{Ext}\,[Y|\Lambda]$

(iv) Ext $[Y|\Lambda] \subset$ Ext $[Y+\Lambda|\Lambda]$ if Λ contains no nontrivial subspace.

Remark 1.7. (ii) of Theorem 1.1 coincides with our intuition that that larger is the domination cone the smaller is the set of all nondominated solutions. This property is very important in our later discussion.

Observe, if L is the maximum linear subspace contained in Λ, then by the decomposition theorem, we can write $\Lambda = L \oplus \Lambda^{\perp}$, where $\Lambda^{\perp} = \Lambda \cap L^{\perp}$ ($L^{\perp}$ is the orthogonal space of L). Note that if $L = \{0\}$, then $\Lambda = \Lambda^{\perp}$. Also note that $\Lambda^{\perp}$ is uniquely determined for each Λ.

Theorem 1.2. (Theorem 4.1. of [36])

A necessary and sufficient condition for $y_0 \in$ Ext $[Y|\Lambda]$ is

(i) $Y \cap (y_0 + L) = \{y_0\}$ and

(ii) $y_0^{\perp} \in$ Ext $[Y^{\perp}|\Lambda^{\perp}]$

where L is the maximum linear space contained in Λ, and $y_0^{\perp}$, $Y^{\perp}$, $\Lambda^{\perp}$ are the projections of y_0, Y and Λ into $L^{\perp}$ respectively.

Remark 1.8. Whenever Λ contains a nontrivial subspace, troublesome complications often arise in our analysis. For examples of this see [36]. Theorem 1.2. implies that we can avoid such pathological behavior by working with $Y^{\perp}$ and $\Lambda^{\perp}$.

Corollary 1.1.

Suppose that $\Lambda^{\perp} \neq \{0\}$. Then

(i) $y_0 \in$ Ext $[Y|\Lambda]$ implies that y_0 is a boundary point of $Y + \Lambda$.

(ii) Ext $[Y|\Lambda] = \phi$ whenever $Y + \Lambda = R^{\ell}$. (This corollary comes from Corollary 4.1-4.2 of [36].)

Theorem 1.3.

Suppose that D(y) is convex for each $y \in Y$. Then

$N_Y(D(\cdot)) \subset \text{Ext}\,[Y|\Lambda^0]$, where

$\Lambda^0 = \cap\,\{D(y)|y \in Y\}$ (i.e., the intersection of all D(y), $y \in Y$).

Assume further that each D(y) contains no non-trivial subspace. Let $Y_y = Y \cap (y + D(y))$. Then

(i) $\{y_0\} = \text{Ext}\,[Y_{y_0}|D(y_0)]$

(ii) If $N_0 \subset N_Y(D(\cdot))$ and $Y_{N_0} = \cup\{Y_y|y \in N_0\}$ then $N_0 \subset \text{Ext}\,[Y_{N_0}|\Lambda_{N_0}]$ where $\Lambda_{N_0} = \cap\,\{D(y))|y \in N_0\}$

(iii) if $Y = \cup\,\{Y_y|y \in N_Y(D(\cdot))\}$ then $N_Y(D(\cdot)) \subset \text{Ext}\,[Y|\Lambda_N]$ where $\Lambda_N = \cap\,\{D(y)|y \in N_Y(D(\cdot)\}$

(The above theorem comes from Lemma 5.1. of [36])

Definition 1.4. Suppose that $\Lambda^0 = \cap\,\{D(y)|y \in Y\} \neq \{0\}$. For $n = 0, 1, 2, \ldots$, we construct two sequences $\{Y^n\}$ and $\{\Lambda^n\}$ as follows, $Y^0 = Y$ and

$Y^{n+1} = \text{Ext}\,[Y^n|\Lambda^n]$ with

$\Lambda^n = \cap\,\{D(y)|y \in Y^n\}$

Theorem 1.4. (Lemma 5.2. of [36])

(i) $Y^n \supset Y^{n+1}$ and each $Y^n \supset N_Y\,(D(\cdot))$.

(ii) $\{Y^n\}$ has the limit $\tilde{Y} = \cap\,\{Y^n|n = 0, 1, \ldots\}$ and $\tilde{Y} \supset N_Y(D(\cdot))$.

Remark 1.9. Theorem 1.3. can be used to estimate a domination cone while Theorem 1.4. can be used as an iterative procedure to locate $N_Y(D(\cdot))$.

Because cone extreme points play a vital role in applications and in computing nondominated solutions, in the next two sections we shall focus our attention on basic necessary and sufficient conditions for a point in Y to be a cone extreme point.

2. ADDITIVE WEIGHTS MAXIMIZATION, CONE CONVEXITY AND CONE EXTREME POINTS

Definition 2.1. The polar cone of Λ is defined by $\Lambda^* = \{\lambda | \lambda \cdot d \leq 0$ for all $d \in \Lambda\}$.

Definition 2.2. A cone (not necessarily closed or convex) is said to be acute if there is an open half space H such that $\bar{\Lambda} \subset H \cup \{0\}$ (where $\bar{\Lambda}$ is the closure of Λ).

It is know that[†] $\text{Int}\, \Lambda^* \neq \phi$ if and only if Λ is acute, furthermore we can write $\text{Int}\, \Lambda^* = \{\lambda | \lambda \cdot d < 0$ for all nonzero $d \in \Lambda\}$. (See Theorem 2.1. of [36].)

Definition 2.3. Let λ be a nonzero vector. Define $Y^0(\lambda) = \{y^0 \in Y | \lambda \cdot y^0 = \sup \lambda \cdot y, y \in Y\}$. Note that $Y^0(\lambda)$ is the set of all maximal points of $\lambda \cdot y$ over Y.

Theorem 2.1.

(i) If $\lambda \in \text{Int}\, \Lambda^*$, then $Y^0(\lambda) \subset \text{Ext}\, [Y|\Lambda]$. Thus $\cup\{Y^0(\lambda) | \lambda \in \text{Int}\, \Lambda^*\} \subset \text{Ext}\, [Y|\Lambda]$.

(ii) Suppose that Λ^* is acute. Let $\Lambda_0 = \{0\} \cup \text{Int}\, \Lambda$. Then $\lambda \in \Lambda^*$, $\lambda \neq 0$, implies that $Y^0(\lambda) \subset \text{Ext}\, [Y|\Lambda_0]$. Thus $\cup\{Y^0(\lambda) | \lambda \in \Lambda^*, \lambda \neq 0\} \subset \text{Ext}\, [Y|\Lambda_0]$.

(iii) If $\lambda \in \Lambda^*$ and $Y^0(\lambda)$ contains only one point, then $Y^0(\lambda) \subset \text{Ext}\, [Y|\Lambda]$.

((i) and (iii) come from Lemma 4.4 of [36], (ii) can be proven easily by using the assumption on acute cone.)

[†] $\text{Int}\, \Lambda^*$ denotes the interior of Λ^*.

Remark 2.1. Note that if Int $\Lambda^* \neq \phi$ then $(\text{Int } \Lambda^*)^* = (\Lambda^*)^* = \bar{\Lambda}$ (the closure of Λ). Thus according to (i)-(ii) of Theorem 2.1., if the set of all possible weight vectors λ is bounded by the interior of a convex cone Λ^* or by a closed convex cone Λ^*, then each optimal solution will be a cone extreme point with respect to the cone $\bar{\Lambda} = (\Lambda^*)^*$ or $\Lambda_0 = \{0\} \cup \text{Int } (\Lambda^*)^*$. In this way, we see that by specifying a set of possible weight vectors and using the additive weight method, we implicitly induce a domination cone so that the candidates for the final decision will be the related cone extreme points. Note that specifying a set of possible weight vectors and using the additive weight method involves a more strict assumption than the induced domination cone does. This is because the former produces less candidates for final decision than the latter does. (See Example 1.3.)

As a consequence of Theorem 2.1., we have

Corollary 2.1. Ext $[Y|\Lambda] \neq \phi$ if one of the following conditions holds:

(i) there is $\lambda \in \Lambda^*$ so that $Y^0(\lambda)$ contains only one point,

(ii) Y is compact and Λ is acute.

Definition 2.4. Given a set S and a cone Λ in R^ℓ, we say that S is Λ -convex ("cone-convex") if and only if

$$S + \Lambda = \{s + \lambda | s \in S, \lambda \in \Lambda\} \text{ is a convex set.}$$

Example 2.1. Given Λ be specified as in Figure 2.0, S_1 in Figure 2.1. is Λ-convex, but S_2 in Figure 2.2 is not Λ-convex.

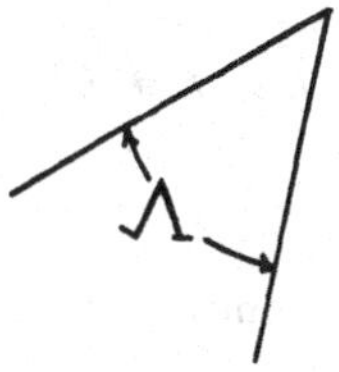

Figure 2.0

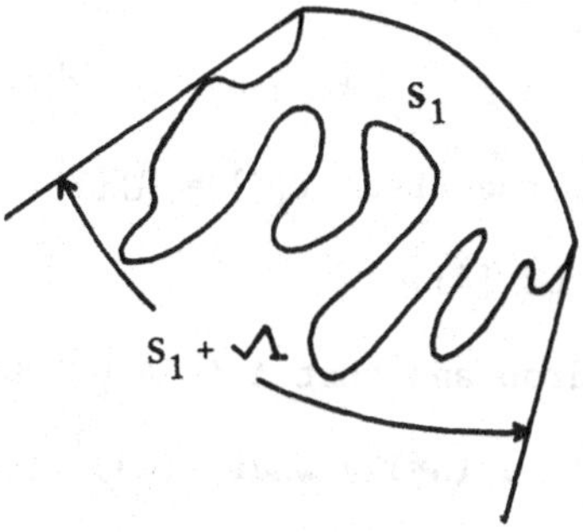

Figure 2.1

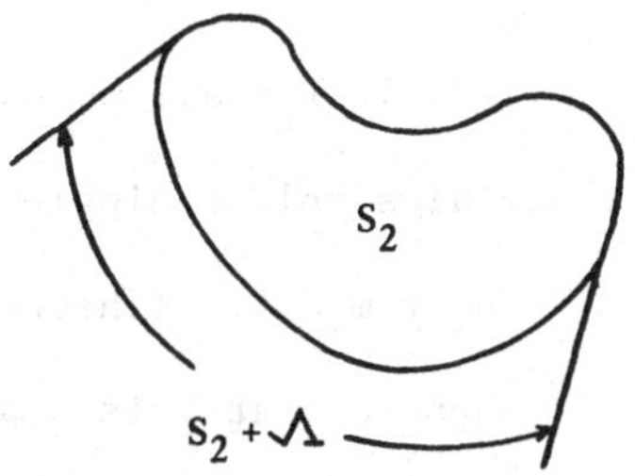

Figure 2.2

A cone which is a closed polyhedron is called a <u>polyhedral cone</u>. If Λ is a polyhedral cone, then Λ^* is also a polyhedral cone and there are a finite number of vectors $\{H^1, \ldots, H^q\}$ so that $\Lambda^* = \{\sum_{i=1}^{m} a_i H^i \mid a_i \in R^1, a_i \geqq 0\}$. $\{H^1, \ldots, H^q\}$ will be called a generator (or a generating set) for Λ^*.

<u>Theorem 2.2.</u> (Lemma 3.1. and Corollary 3.4. of [36])

(i) Let Λ_1, Λ_2 be two convex cones in R^m such that $\Lambda_1 \subset \Lambda_2$. Then Y is Λ_1-convex implies that Y is Λ_2-convex.

(ii) Let Λ be a polyhedral cone with $\{H^1, \ldots, H^q\}$ as a generator for Λ^*. Suppose that X is a convex set and that each $H^j \cdot f(x)$ is a concave function over X. Then $Y = f[X]$ is Λ-convex.

<u>Example 2.2.</u> In SIP (see Example 1.1), clearly f_1 is a concave function. Suppose that f_2 is linear. Then Y is Λ-convex for any Λ containing $U = \{a(-1,0) \mid a \in R^1, a \geqq 0\}$, because Y is U-convex. (Note that $\{(0,1), (0,-1), (1,0)\}$ is a generator for U^*. As a consequence of Theorem 2.2, Y is U-convex.)

<u>Theorem 2.3.</u> (Corollary 4.7 of [36])

Suppose Y is Λ-convex and $\Lambda^\perp \neq \{0\}$. Then

(i) $\cup \{Y^0(\lambda) \mid \lambda \in \text{Int } \Lambda^*\} \subset \text{Ext } [Y|\Lambda] \subset \cup \{Y^0(\lambda) \mid \lambda \in \Lambda^*, \lambda \neq 0\}$

(ii) If for all boundary points of Λ^* except 0, $Y^0(\lambda)$ is either empty or contains only a single point, then $\text{Ext}\,[Y|\Lambda] = \cup\{Y^0(\lambda)|\lambda \in \Lambda^*, \lambda \neq 0\}$

Theorem 2.4. (Theorem 2.5 of [41])

Suppose that Y is a polyhedron and that Λ is a polyhedral cone. Then

(i) $\text{Ext}\,[Y|\Lambda] \subset \cup\,\{Y^0(\lambda)|\lambda \in (\Lambda^*)^I\}$ where $(\Lambda^*)^I$ is the relative interior of Λ^*.

(ii) If $\text{Int}\,\Lambda^* \neq \phi$, then $\text{Ext}\,[Y|\Lambda] = \cup\{Y^0(\lambda)|\lambda \in \text{Int}\,\Lambda^*\}$.

Remark 2.2. Theorem 2.3 is a generalization of a decision theory theorem that every admissible strategy is Bayes, and that every strategy which is Bayes against positive weights is admissible. Theorem 2.4 states an important property which is enjoyed by linear multicriteria decision problems. The theorem in fact asserts that in statistical decision problems which have a finite number of states of uncertainty and a finite number of actions from which to choose every admissible strategy is a Bayes against positive weights and conversely, every Bayesian strategy against positive weights is admissible.

Theorems 2.3 and 2.4 also have an important implication for the construction of utility functions. They give conditions which imply that a utility function can be approximated by using the additive weight method. They also state the bounds for the weights to be used. In order to demonstrate this point, we introduce

Definition 2.5. A function U(y) over Y is said to be decreasing in the direction $d \neq 0$ if $U(y + \lambda d) < U(y)$ whenever $y \in Y$, $y + \lambda d \in Y$ and $\lambda > 0$. We say that U(y) is decreasing in a convex cone Λ if it is decreasing in each direction represented by the non-zero elements of Λ.

Theorem 2.5.

Suppose that U(y) is decreasing in Λ over Y. Suppose also that Y is Λ-convex and $\Lambda^{\perp} \neq \{0\}$. Then there is $\lambda \in \Lambda^*$, $\lambda \neq 0$, such that the maximum point of U(y) is also a maximum point of $\lambda \cdot y$ over Y. When both Y and Λ are polyhedral, λ can be selected from $(\Lambda^*)^I$.

Proof

By assumption, the maximum point of U(y) over Y must be a Λ-extreme point. Our assertions then follow from Theorems 2.3 and 2.4.

Example 2.3. In the SIP, suppose that f_2 (the expected return) is linear. (See Example 2.2.) Suppose that the investor is risk averse. That is if $f_2(x) = f_2(x')$ and $f_1(x) > f_1(x')$ then f(x) is preferred to f(x'). Also if $f_1(x) = f_1(x')$ and $f_2(x) > f_2(x')$ then f(x) is preferred to f(x'). Since the higher is f_2 the more preferred will it be, we may assume that the utility function, if it exists, is decreasing in $\Lambda^{\leqq}$. Thus the maximization of the utility function may be approximated by the maximization of an additive weight function $\lambda_1 f_1 + \lambda_2 f_2$ with $\lambda_1, \lambda_2 \geqq 0$ and λ_1, λ_2 not both zero.

3. SATISFICING, ADDITIVE WEIGHT MAXIMIZATION AND CONE EXTREME POINTS

In the previous section, we described the main characteristics of Ext $[Y|\Lambda]$ whenever Y is Λ'-convex for some $\Lambda' \subset \Lambda$. In this section cone convexity assumption on Y will be removed. We shall show that cone extreme points enjoy the unique maximization of an additive weight objective function subject to a set of additive weight constraints. The objectives and the constraints are interchangeable. We shall present two basic formulations.

Throughout this section, unless otherwise specified, we shall assume that Λ is a polyhedral cone with $\{H^1,\ldots, H^q\}$ as a generator for Λ^*. For a generalization of the results to be described, we refer to [36].

Definition 3.1. Given $y^0 \varepsilon Y$, we define $Y_j(y^0) = \{Y \varepsilon Y | H^k\cdot y \geqq H^k\cdot y^0, k = 1,\ldots, q, k \neq j\}$.

Note that $Y_j(y^0)$ is the set of all y in Y which makes $H^k\cdot y$ at least as large as $H^k\cdot y^0$ for all k =],..., q except j.

Theorem 3.1. (Theorem 4.2 of [36])

$y^0 \varepsilon$ Ext $[Y|\Lambda]$ if and only if for any arbitrary j = 1,..., q, y^0 uniquely maximizes $H^j\cdot y$ over all $y \varepsilon Y_j(y^0)$.

Remark 3.1. $H^j\cdot y = H^j\cdot f(x)$ is an additive weight form. Since j is arbitrary, any constraint and the objective function are interchangeable. Note that "uniqueness" is an important property which reveals that in using ordinary mathematical programming we have to pay special attention to alternative "optimal" solutions; otherwise, we may obtain a "dominated" solution.

Remark 3.2. Theorem 3.1 can be used to verify whether a particular point is nondominated or not. However, using it to locate the set of nondominated solutions is computationally inconvenient. Thus we shall discuss a second procedure derived from Theorem 3.1.

Definition 3.2. Let r(j) be the vector in R^{q-1} representing $\{r_k \varepsilon R^1 | k \neq j, k = 1,\ldots, q\}$. We define

$$Y(r(j)) = \{y \varepsilon Y | H^k\cdot y \geqq r_k, k \neq j, k = 1,\ldots, q\}.$$

Note that Y(r(j)) is the set of points in Y which satisfy $H^k\cdot y \geqq r_k$, $k \neq j$, k = 1,..., q. Note, r_k may be regarded as a satisficing level for $H^k\cdot y$.

Theorem 3.2. (Theorem 4.3 of [36])

$y^0 \in \text{Ext}\,[Y|\Lambda]$ if and only if for any arbitrary $j = 1,\dots, q$ there is $r(j)$ such that y^0 uniquely maximizes $H^j \cdot y$ over $Y(r(j))$.

Remark 3.3. Note that $(\Lambda^{\leqq})^* = \Lambda^{\geqq} = \{d \in R^{\ell} | d \geqq 0\}$. Thus the set $\{e^j | j = 1,\dots,\ell\}$, where e^j is the jth column of the $\ell \times \ell$ identity matrix, is a generator for $\Lambda^{\leqq}$. Note that $e^j \cdot y = y_j$ (the jth component of y). With this observation and Theorems 3.1 - 3.2, we have the following results.

Theorem 3.3.

$y^0 \in \text{Ext}\,[Y|\Lambda^{\leqq}]$ (i.e., y^0 is Pareto optimal or efficient) if and only if for any arbitrary $j = 1,\dots, \ell$, y^0 uniquely maximizes y_j over all $y \in \{y \in Y | y_k \geqq y_k^0,\ k = 1,\dots, \ell,\ k \neq j\}$.

Theorem 3.4.

$y^0 \in \text{Ext}\,[Y|\Lambda^{\leqq}]$ if and only if for any arbitrary $j = 1,\dots, \ell$ there are ℓ-1 constants $\{r_k | k = 1,\dots, \ell,\ k \neq j\}$ such that y^0 uniquely maximizes y_j for all $y \in \{y \in Y | y_k \geqq r_k,\ k = 1,\dots, \ell,\ k \neq j\}$.

Example 3.1.

In SIP suppose that $\Lambda = \left\{(d_1, d_2) \middle| \begin{array}{l} d_1 + 2d_2 \leqq 0 \\ 2d_1 + d_2 \leqq 0 \end{array}\right\}$. Then

$\Lambda^* = \left\{(\lambda_1, \lambda_2) \cdot \begin{bmatrix} 1 & 2 \\ 2 & 1 \end{bmatrix} \middle| \lambda_1, \lambda_2 \geqq 0\right\}$. (See Figure 1.) That is, (1,2) and (2,1) form a generator for Λ^*. Suppose that Y is as in Figure 3. Then $y^0 \in \text{Ext}\,[Y|\Lambda]$ because $(1,2)\cdot(y_1,y_2)$ has the unique maximum at y^0 over $Y_2(y^0)$. Note that y^* is not a Λ-extreme point because it does not maximize $(1,2)\cdot(y_1,y_2)$ over $Y_2(y^*)$.

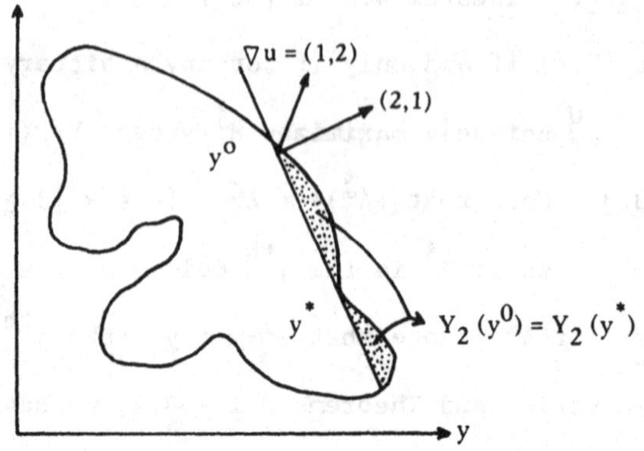

Figure 3

4. NONDOMINATED SOLUTIONS IN THE DECISION SPACE

In this section we shall derive conditions for nondominated solutions in the decision space. For simplicity we shall focus on one-stage decision problems. The extensions of our results to dynamic cases are straight forward. We shall refer to [40] for such extensions.

In order to make our presentation more precise, let us define the decision space by

$$X = \{x \in R^n | g_i(x) \leqq 0, \ i = 1, \ldots, m\}$$

or $$X = \{x \in R^n | g(x) \leqq 0\} \text{ with } g = (g_1, \ldots, g_m). \tag{4.1}$$

Given a convex cone Λ in R^m, we define the set of all Λ-extreme points in the decision space by

$$X^0(\Lambda) = \{x \in X | f(x) \in \text{Ext } [Y|\Lambda]\}. \tag{4.2}$$

Thus,

$$\text{Ext } [Y|\Lambda] = f[X^0(\Lambda)] = \{f(x) | x \in X^0(\Lambda)\}.$$

Now by applying the results in Section 2 and 3 we can derive the conditions for nondominated decisions. For instance if we apply Theorem 3.1, we will have

Theorem 4.1.

Suppose Λ is a polyhedral cone with Λ^* generated by $\{H^1,\ldots, H^q\}$. Then

(i) A necessary condition for $x^0 \in X^0(\Lambda)$ is that for any arbitrary $j \in \{1,2,\ldots,q\}$ there are q-1 real numbers $\{r_k | k \neq j, k = 1,\ldots, q\}$ such that x^0 maximizes $H^j\cdot f(x)$ for $x \in \{x \in X | H^k\cdot f(x) \geqq r_k, k = 1,\ldots, q, k \neq j\}$,

(ii) If x^0 is the unique maximizing decision for the problem stated in (i), then $x^0 \in X^0(\Lambda)$.

Clearly by applying the different results in Section 2-3, we will obtain a different set of conditions for $x^0 \in X^0(\Lambda)$. We shall leave these applications to the reader.

From now on we shall assume the following

Assumption 4.1

(i) Λ is specified as in Theorem 4.1 and $\Lambda^\perp \neq \{0\}$;

(ii) $H^k\cdot f(x)$ is concave and differentiable over X for each $k = 1,\ldots,q$;

(iii) $g(x)$ is quasiconvex and differentiable over X and Kuhn-Tucker constraint qualification† is satisfied at each point of X.

Note that according to (ii) of Theorem 2.2, Y is Λ-convex. Thus Theorem 2.3 can be applied. Define

$$X_0(\lambda) = \{x^0 \in X | x_0 \text{ maximizes } \lambda\cdot f(x) \text{ over } X\} \quad (4.3)$$

$$X_A(\Lambda) = \cup\{X_0(\lambda) | \lambda \in \Lambda^*, \lambda \neq 0\} \quad (4.4)$$

$$X_A^I(\Lambda) = \cup\{X_0(\lambda) | \lambda \in \text{Int } \Lambda^*\} \quad (4.5)$$

---

†See [22] for definition.

According to Theorem 2.3 and (4.2)-(4.5) we have

$$X_A^I(\Lambda) \subset X^0(\Lambda) \subset X_A(\Lambda) \qquad (4.6)$$

From standard Kuhn-Tucker conditions, we have: (Recall that Assumption 4.1 is assumed)

Lemma 4.1

$x^0 \in X_o(\lambda)$ iff there is $\mu \in R^r$ such that

$$\lambda \cdot \nabla f(x^0) - \mu \cdot \nabla g(x^0) = 0 \qquad (4.7)$$

$$\mu \cdot g(x^0) = 0 \qquad (4.8)$$

$$g(x^0) \leqq 0 \qquad (4.9)$$

$$\mu \geqq 0 \qquad (4.10)$$

Let $I(x^0) = \{i | g_i(x^0) = 0\}$ and let $\mu_{I(x^0)}$ and $g_{I(x^0)}(x^0)$ be the vectors derived from μ and $g(x^0)$ by deleting all components of μ and $g(x^0)$ which are not in $I(x_0)$. Define

$$F(x^0,\Lambda) = \{\lambda \cdot \nabla f(x^0) | \lambda \in \Lambda^*, \lambda \neq 0\} \qquad (4.11)$$

$$F^I(x^0,\Lambda) = \{\lambda \cdot \nabla f(x^0) | \lambda \in \text{Int } \Lambda^*\} \qquad (4.12)$$

$$G(x^0) = \{\mu_{I(x^0)} \cdot \nabla g_{I(x^0)}(x^0) | \mu_{I(x_0)} \geqq 0\} \qquad (4.13)$$

with the understanding that $G(x^0) = \{0\}$ if $I(x^0) = \emptyset$.

From Lemma 4.1 one obtains

Theorem 4.2.

Suppose Assumption 4.1 is satisfied. Then

(i) $x^0 \in X_A(\Lambda)$ if and only if $x^0 \in X$ and $F(x^0,\Lambda) \cap G(x^0) \neq \emptyset$

(ii) $x^0 \in X_A^I(\Lambda)$ if and only if $x^0 \in X$ and $F^I(x^0,\Lambda) \cap G(x^0) \neq \emptyset$.

Remark 4.1. Similar results for Pareto optimal solutions can be easily derived. We shall not stop to do so. Note that $F(x^0,\Lambda)$, $F^I(x^0,\Lambda)$ and $G(x^0)$ are convex cones which are uniquely determined for each x^0. When g and f are linear or quadratic, such cones will be constant or vary linearly with x^0. Then, the condition in Theorem 4.2 can be verified by verifying that a system of linear inequalities has solution.

Let $M = \{1,2,\ldots,m\}$ and $\mathcal{M} = \{J | J \subset M\}$. For $J \in \mathcal{M}$ define $X_J = \{x | g_J(x) \leqq 0\}$ where $g_J(x)$ is derived from $g(x)$ by deleting all components of $g(x)$ except those in J. Let $X_J^0(\Lambda)$ be the set of all Λ-extreme points in X_J. Let $X_{0J}(\lambda) = \{x^0 \in X | x_0$ maximizes $\lambda \cdot f(x)$ over $X_J\}$, and let $X_{JA}^I(\Lambda)$ and $X_{JA}(\Lambda)$ be defined using (4.4) and (4.5) with $X_{0J}(\lambda)$ in place of $X_0(\lambda)$.

Theorem 4.3.

(i) $X_A(\Lambda) = \bigcup_{J \in \mathcal{M}} (X_{JA}(\Lambda) \cap X)$

(ii) $X_A^I(\Lambda) = \bigcup_{J \in \mathcal{M}} (X_{JA}^I(\Lambda) \cap X)$

Let the number of elements in J be denoted by [J] and let $\mathcal{J}^k = \{J \in \mathcal{M} | [J] = k\}$, $k = 0,1,\ldots,m$. For each $J \in \mathcal{M}$, define $\tilde{X}_J = \{x | g_J(x) = 0\}$. We have the following useful decomposition theorem.

Theorem 4.4.

If Assumption 4.1 is satisfied, then

(i) $X_A(\Lambda) = (X_{\phi A}(\Lambda) \cap X) \cup \left(\bigcup_{k=1}^{m} \bigcup_{J \in \mathcal{J}^k} [X_{JA}(\Lambda) \cap \tilde{X}_J \cap X]\right).$

(ii) $X_A^I(\Lambda) = (X_{\phi A}^I(\Lambda) \cap X) \cup \left(\bigcup_{k=1}^{m} \bigcup_{J \in \mathcal{J}^k} [X_{JA}^I(\Lambda) \cap \tilde{X}_J \cap X]\right).$

Remark 4.2. The set $X_{JA}(\Lambda) \cap \tilde{X}_J \cap X$ could be located by first finding those points $X_{JA}(\Lambda)$ on $\tilde{X}_J$, that is $X_{JA}(\Lambda) \cap \tilde{X}_J$, and then discarding those points of $X_{JA}(\Lambda) \cap \tilde{X}_J$ which violate the constraints $g_i(x) \leqq 0$, $i \notin J$. Thus, Theorem 4.4 could be used systematically to locate X_{JA}, X_{JA}^I and $X^0(\Lambda)$.

In addition, it could be used to produce for all $x^0 \in X^0(\Lambda)$, $\Lambda^*(x_0) = \{\lambda | \lambda \in \Lambda^*, \lambda \neq 0$ such that $x^0 \in X^0(\lambda)\}$. Thus if $\lambda \in \Lambda^*(x_0)$ then x^0 maximizes $\lambda \cdot f(x)$ over X. Of course, $\Lambda^*(x^0)$ is extremely useful because it specified properties of x^0 and thus aids the final decision making. An example along this line can be found in [8,36].

So far we have only discussed nondominated decisions with respect to a constant domination cone. To obtain the conditions for nondominated decisions with respect to general domination structures, one can apply Theorem 1.3 and 1.4. We shall not develop this procedure here. For a detailed discussion we refer the reader to [8,36].

5. NONDOMINATED SOLUTIONS IN LINEAR CASES AND A MULTICRITERIA SIMPLEX METHOD

When both F(x) and g(x) are linear, the nondominated solutions enjoy some special properties. As a consequence the computation of the set of all nondominated solutions is greatly simplified in linear cases. In this section we shall discuss such special properties and introduce a multicriteria simplex method to facilitate the computation of the set of nondominated solutions.

In order to simplify the presentation, we shall use the following notation:

$$X = \{x \in R^n | \; Ax \leqq b, \quad x \geqq 0\} \tag{5.1}$$

where A is of order m x n,

$$Y = \{Cx | \; x \in X\} \tag{5.2}$$

where C is of order ℓ x n so that $C^k \cdot x$, $k = 1, \ldots, \ell$, (C^k is the kth row of C), is the kth criterion. Given a matrix C, we shall use C^k and C_j

to indicate the kth row and the jth column of C respectively.

We shall assume that for all $y \in Y$, $D(y) = \Lambda$, a fixed convex domination cone; we use "N-point" and "D-point" to denote a nondominated point and a dominated point respectively. The sets of all N-points and all D-points will be denoted by N and D respectively. Besides Theorem 2.4, we have

Theorem 5.1. (Lemma 2.1, Theorem 2.1-2.2 of [41].)

(i) Suppose $x^1, x^2 \in X$ and $x^1 \in D$. Then $[x^1, x^2) \subset D$ where $[x^1, x^2) = \{\alpha x^1 + (1-\alpha)x^2 | 0 < \alpha \leqq 1\}$.

(ii) D is a convex set

(iii) Let K be an arbitrary convex subset of X. Then $K^I \subset D$ whenever $K^I \cap D \neq \emptyset$; and $\bar{K} \subset N$ whenever $K^I \cap N \neq \emptyset$. (Recall that K^I and $\bar{K}$ represent respectively the relative interior and the closure of K.)

Since every polyhedral cone can be linearly transformed into a form of $\Lambda^{\leqq}$ (possibly with a different dimension) [36], we shall from now on assume $\Lambda^{\leqq}$ as our domination cone. Also we shall assume that X is compact. This assumption will greatly simplify our presentation. We note that our main results can be extended to the case where X is unbounded.

Since X is compact, the set of its extreme points is finite and will be denoted by X_{ex}. The set of all nondominated extreme points will be denoted by N_{ex}. That is

$$N_{ex} = N \cap X_{ex}.$$

Theorem 5.2.†

$N \subset H(N_{ex})$ (the convex hull of N_{ex}).

---

†When X is unbounded, N is contained in the sum of $H(N_{ex})$ and the nonnegative cone generated by the nondominated extreme rays.

Remark 5.1. Theorem 5.1 says that a face of X is either dominated or nondominated. The checking of the dominance can be performed at any relative interior point of the face. In particular any face which contains a D-point must be dominated (of course we mean its relative interior).

Remark 5.2. Theorem 5.2 suggests that we should first find N_{ex} and then use it to generate the set N. Clearly we will save a great deal of computation if N_{ex} can be easily found and N can be easily generated from N_{ex}.

We first introduce a multicriteria simplex method to locate N_{ex}. The method can be coded for computer computation.

The main difference between the ordinary simplex method and the multicriteria simplex method is in the rows which are associated with the criteria. The former has only one row associated with the criterion, while the latter has ℓ rows associated with the criteria.

Without loss of generality, we shall assume that $b \geqq 0$ in (5.1). By adding slack variables, the decision space could be defined by the set of all $x \in R^{m+n}, x \geqq 0$ and

$$(A, I_{mxm})\ x = b. \tag{5.3}$$

Our new C becomes (C, O_{mxm}). (5.4)

Given a basis B which is associated with columns $J = j_1, j_2, \ldots, j_m$, we shall denote the remaining submatrix and columns with respect to (5.3) by B' and J' respectively.

We construct a multicriteria simplex tableau as Tableau 1 (for simplicity, we have rearranged the indices so that J appears in the first m columns).

r	BASIS	x_1	...	x_m	x_{m+1}	...	x_j	...	x_{m+n}	x
1	x_1	1	...	0	y_{1m+1}	...	y_{1j}	...	y_{1m+n}	y_{10}
⋮	⋮	⋮	⋮	⋮	⋮	⋮	⋮	⋮	⋮	⋮
m	x_m	0	...	1	y_{mm+1}	...	y_{mj}	...	y_{mm+n}	y_{m0}
		0	...	0	z^1_{m+1}	...	z^1_j	...	z^1_{m+n}	v^1
		⋮	⋮	⋮	⋮	⋮	⋮	⋮	⋮	⋮
		0	...	0	z^ℓ_{m+1}	...	z^ℓ_j	...	z^ℓ_{m+n}	v^ℓ

Tableau 1

Where $Y = \{y_{ij}\}_{\substack{i=1,\dots,m\\ j=1,\dots,m+n}} = B^{-1}[A, I_{mxm}] = B^{-1}(B, B')$

$$= (I, B^{-1}B') \tag{5.5}$$

$$Z = \{z^k_j\}_{\substack{k=1,\dots,\ell\\ j=1,\dots,m+n}} = C_BY - C = C_B(I, B^{-1}B') - (C_B, C_B')$$

$$= (0, C_BB^{-1}B' - C_B') \tag{5.6}$$

(where C_B and C_B' are the cost coefficients associated with B and B' respectively),

$$y_0 = (y_{10}, \dots, y_{m0})^T = B^{-1}b \tag{5.7}$$

and

$$V = (v^1, \dots, v^\ell) = C_BB^{-1}b \tag{5.8}$$

Let us define M as

$$M = \begin{bmatrix} Y \\ Z \end{bmatrix}_{(m+\ell) \times (m+n)} \tag{5.9}$$

Observe that M enjoys the following properties,

(i) the submatrix $\{y_j | j \varepsilon J\}$, when its rows are properly permutated, forms the identity matrix of order m x m, (5.10)

(ii) the submatrix $\{Z_j | j \varepsilon J\}$ is a zero matrix of order ℓ x m. (5.11)

Obtaining an initial tableau for our problem is a simple application of Gaussian elimination technique. For each nonbasic column $j \varepsilon J'$, we shall define θ_j as follows

$$\theta_j = \frac{y_{p0}}{y_{pj}} = \min_r \{ \frac{y_{r0}}{y_{rj}} | y_{rj} > 0 \} \tag{5.12}$$

Note that θ_j is well defined, because X is compact (thus unbounded solutions are impossible).

In our iterations of the multicriteria simplex method, by introducing column $j \varepsilon J'$ into the basis, we convert M_j into E_p in the next tableau, where E_p is the pth column of the identity matrix of order $m + \ell$ and p is such that (p,j) is the pivot element. At each such iteration M enjoys the properties (5.10)-(5.11) and Y,Z can be easily computed.

Remark 5.3. Observe that the multicriteria simplex method is different from ordinary simplex method only in Z, not in Y. One can think that Z is associated with ℓ-criteria which we want to "maximize" over the same feasible set X. In fact, the Z^k, $k = 1, \ldots, \ell$, is associated with objective function $C^k x$. If we are only interested in the kth criterion, then Z^k is the row of optimality condition in ordinary simplex method. Thus given a basis J, if $Z^k \geqq 0$, then x(J), the basic feasible solution of J, is an optimal solution for the kth criterion.

Remark 5.4. Given a basis J and $j \in J'$ by introducing j^{th} column into the basis we produce an adjacent basis J_1. From the simplex method and Remark 5.3, we see that the values of the objective functions increase by $-\theta_j Z_j$. That is $V(J_1) - V(J) = -\theta_j Z_j$, where $V(J) = (v^1, \ldots, v^{\ell})$ at the basis J. This observation yields:

Theorem 5.3

Given a basis J_o

(i) If there is i, $1 \leq i \leq \ell$, so that each $z^i_j > 0$ for all $j \in J'_o$ then $x(J_o) \in N_{ex}$ (because $x(J_o)$ uniquely maximizes $C^i x$ over X).

(ii) If there is $j \in j'_o$ so that $\theta_j Z_j \leqq 0$, $\theta_j Z_j \neq 0$, then $x(J_o) \in D$.

(iii) If there is $j \in j'_o$ so that $\theta_j Z_j \geqq 0$, $\theta_j Z_j \neq 0$, then $x(J_1) \in D$, where J_1 is the new basis obtained by introducing the jth column into the basis.

(iv) Let $j, k \in J'_o$ with $\theta_j, \theta_k < \infty$ and, J_j and J_k be the new basis obtained by introducing respectively the jth and k^{th} column into the basis. Suppose that $\theta_j Z_j \geqq \theta_k Z_k$, $\theta_j Z_j \neq \theta_k Z_k$. Then $x(J_j) \in D$.

Now let us consider the problem of maximizing λCx over X. Let $z(\lambda) = (z_{11}, \ldots, z_{1m+n})$ be the row associated with the optimality condition in the simplex method. Then given a basis B with C_B, C'_B and Y defined as in (5.5)-(5.6), we see that

$$z(\lambda) = (0, \lambda C_B B^{-1} B' - \lambda C'_B) = \lambda(0, C_B B^{-1} B' - C'_B)$$

$$= \lambda(C_B Y - C) = \lambda Z \tag{5.13}$$

Thus we have -

Theorem 5.4

For every basic feasible solution, $z(\lambda) = \lambda Z$.

Theorem 5.5

Given a feasible basis J, let Z(J) be the matrix Z associated with J. Then the basic feasible solution x(J) maximizes λCx over X for all

$$\lambda \in \Lambda(J) = \{\lambda \mid \lambda Z(J) \geqq 0\}. \tag{5.14}$$

Remark 5.5. Observe that $\Lambda(J)$ is a polyhedron and $0 \in \Lambda(J)$. Given x(J), $\Lambda(J)$ is its associated set of optimal weights.

According to Theorem 2.4 and 5.5, we have

Theorem 5.6

A feasible basic solution x(J) is an N-point if and only if $\Lambda(J) \cap \Lambda^{>} \neq \emptyset$.

Although Theorem 5.6 can be used to test whether a basic feasible point is an N-point, its application involves determining whether a certain system of linear inequalities has a solution. This test is usually not very efficient. We shall derive a method called the nondominance subroutine to perform the test.

Let $x^o = x(J)$ represent a basic feasible solution with basis J. Let $e = (e_1, \ldots, e_\ell)$ and

$$w = \max \sum_{i=1}^{\ell} e_i \text{ over}$$

$$\tilde{X} = \{(x,e) \mid x \in X, Cx - e \geqq Cx^o, e \geqq 0\} \tag{5.15}$$

Theorem 5.7

(i) x^o is an N-point if and only if $w = 0$.

(ii) x^o is a D-point if and only if $w > 0$.

Proof

Observe that $(x^o, 0) \in X$. Thus $w \geqq 0$. It sufficies to show (i). However, (i) is another way to define an N-point with respect to the domination cone $\Lambda^{\leqq}$. Q.E.D.

Observe that finding whether w = 0 or not in Theorem 4.4 does not require too much extra work. In order to see this, let B be the basis associated with x^o or J. The problem of (5.15) in a block simplex tableau can be written

$$\left[\begin{array}{cccc} A_{mxn} & I_{mxm} & 0_{mx\ell} & b_{mx1} \\ C_{\ell xn} & 0_{\ell xm} & -I_{\ell x\ell} & Cx^o \\ 0_{1xn} & 0_{1xm} & -1_{1x\ell} & 0 \end{array}\right] \begin{array}{c} (5.16) \\ (5.17) \\ (5.18) \end{array}$$

where $1_{1x\ell} = (1, 1, \ldots, 1)$.

In the above matrix, the first and second columns are the coefficients associated respectively with the original variables and the added slack variables, the third column is the coefficients associated with the new variable e in (5.15). Note that (5.16) is the constraint that $x \in X$, (5.17) is the constraint that $Cx - e \geqq Cx^o$, that (5.18) corresponds to the objective of (5.15).

We could rewrite (5.16)--(5.18) as follows:

$$\left[\begin{array}{cccc} B^{-1}A & B^{-1} & 0_{mx\ell} & B^{-1}b \\ C_B B^{-1}A-C & C_B B^{-1} & I_{\ell x\ell} & 0_{\ell x1} \\ 1_{1x\ell}\,[C_B B^{-1}A-C] & 1_{1x\ell}[C_B B^{-1}] & 0_{1x\ell} & 0 \end{array}\right] \begin{array}{c} (5.19) \\ (5.20) \\ (5.21) \end{array}$$

Note that (5.19) = B^{-1}(5.16), (5.20) = C_B(5.19)-(5.17) (observe that $C_B B^{-1} b = Cx^o$), and (5.21) = $1_{1x\ell}$[(5.20) + (5.18)].

Observe that (5.19)-(5.21) supply a feasible simplex tableau for Problem (5.15) with the basic feasible solution $(x,e) = (x^0,0)$. Comparing that (5.19)-(5.21) with (5.5)-(5.7), we see that (renumbering the indices of x_j if necessary)

$$\begin{bmatrix} B^{-1}A & B^{-1} \\ C_B B^{-1}A - C & C_B B^{-1} \end{bmatrix} = \begin{bmatrix} Y \\ Z \end{bmatrix} \tag{5.22}$$

From (5.19)-(5.22) we see that to construct a simplex tableau for Problem (5.15) does not require too much extra work. The conditions in Theorem 5.7 could be easily verified. In particular, we have the following sufficiency condition.

Theorem 5.8

Given a basis J, suppose $1_{1x\ell}\, Z \geqq 0$.

Then x(J) is an N_{ex}-point.

Proof

Because the first two blocks of (5.21) are given by $1_{1x\ell}\, Z$, $1_{1x\ell} Z \geqq 0$ implies that (x(J), 0) is an optimal solution to (5.15) with value w = 0. Our assertion follows immediately from Theorem 5.7.

Remark 5.6. In order to use the results of (5.19)-(5.22) and Theorems 5.7 and 5.8, one can append an extra row corresponding to the objective function $1_{1x\ell}C$ to the simplex tableau. Suppose that the condition in Theorem 5.8 is not satisfied. Because of the special structure of (5.19)-(5.21), the problem of (5.15) usually can be simply solved in a few iterations. Note that in the subroutine we usually need to check only a small submatrix of the matrix defined by (5.19)-(5.21). In order to see this, observe that if we start with (5.19)-(5.21) as the initial tableau for Problem (5.15), and if $B^{-1}b > 0$, we can delete all basic columns of M (see (5.9)) and all rows of (5.19) (i.e. those rows associated with Y). At each iteration, we need to consider three possible cases in the remaining submatrix. The first case is that all remaining elements of the row (5.21) are nonnegative. Then Problem (5.15) achieves its maximum with $w = 0$. Thus we have an N-point (by Theorem 5.7). The second case is that at least one remaining element of the row (5.21) is negative and each undeleted element of the associated column is nonpositive. Then we can obtain a feasible solution of Problem (5.15) which yields $\sum_i e_i > 0$ (because the associated $\theta_j > 0$ and thus $w > 0$). Thus from Theorem 5.7, we obtain a D-point. The last case is that at least one remaining element of Row (5.21) is negative and some undeleted element (say i, j) of the associated column is positive. Then the associated[†] θ_j will be zero. We can "pivot" at the element (i,j) to obtain a new tableau. Observe that $B^{-1}b$ will not be changed by the pivoting. Thus the process can be repeated until one of the first two cases occurs.

Remark 5.7. Once J is found to be an N_{ex}-basis, (5.14) can be used to find its related set of optimal weights $\Lambda(J)$ with no extra work from the multi-criteria simplex tableau. Thus our remaining crucial task is to find the set N_{ex} using the multicriteria simplex method.

[†] See (5.12)

Definition 5.1. Let $E = \{x(i) \mid i = 1,\ldots, p\}$ be a set of extreme points of X. We say that E is connected if it contains only one point or if for any two points x(j), x(k) in E, there is a sequence $\{x(i_1),\ldots, x(i_r)\}$ in E so that $x(i_\ell)$ and $x(i_{\ell+1})$, $\ell = 1,\ldots, r-1$, are adjacent and $x(i_1) = x(i_r) = x(k)$.

The proof of the following theorem can be found in [41].

Theorem 5.9

The set N_{ex} is connected.

Remark 5.8. In view of Theorem 5.9, we can construct a connected graph $(\mathcal{E},\mathcal{V})$ for N_{ex}, where $\mathcal{V}$ is the set of all vertices corresponding to N_{ex}, and $\mathcal{E}$ is the set of all arcs in the graph. Given† $x^1, x^2 \in N_{ex}$ the arc $a(x^1,x^2)$ which connects x^1 and x^2 is in $\mathcal{E}$ if and only if x^1 and x^2 are adjacent. With this definition we see that the graph $(\mathcal{E},\mathcal{V})$ is connected.

In order to generate the set N_{ex}, we can first find a basis J_1 for an N_{ex}-point, if $N_{ex} \neq \emptyset$. In view of Remark 5.8, if there is any other N_{ex}-point, we must have an N_{ex}-basis J_2 adjacent to J_1. Thus we could use our results of this section to search for such a J_2. If there is no such J_2, J_1 is the unique N_{ex}-point. Otherwise, we consider all adjacent, but unexplored feasible bases to $\{J_1,J_2\}$ to see if there is any other N_{ex}-basis among them. If there is none, $\{J_1, J_2\}$ represents the set N_{ex}. Otherwise, we add a new N_{ex}-basis to $\{J_1, J_2\}$ and continue with the procedure until the entire set N_{ex} is located.

---

†It is convenient, without confusion, for us to use x^1, x^2 to represent both the associated bases J_1, J_2 and the resulting basic feasible solutions $x(J_1)$, $x(J_2)$.

The method just sketched above can be carried out by computer. For more details of such development and examples, we refer the reader to [41].

We now turn to the problem of generating N from N_{ex}. We shall briefly discuss one procedure. For a detailed development and examples we refer the reader to [41].

One can easily check that usually only a proper subset of $H[N_{ex}]$ is nondominated. The decomposition theorem (Theorem 4.4) supplies an algorithm for the checking of nondominance, while Theorem 5.1 can be used to facilitate the checking. The following observations can be used to speed up the checking:

(i) Our goal is to find a set of nondominated faces $\{F_i\}$ so that each F_i has a maximum dimension (that is, there is no other nondominated face F_k which properly contains F_i) and $N = \bigcup_i \{F_i\}$. Thus our checking can start with the n-dimensional face (that is X), then the (n-1)-dimensional faces and so on.

(ii) F_i is nondominated only when all of its subfaces are nondominated. Thus, once F_i is found to be nondominated, all of its subfaces (no matter what their dimensionality) can be eliminated from checking. Also once we find a D-point (say by the previous multicriteria simplex method), all the faces which contain this D-point must be dominated and can be eliminated from consideration.

(iii) The incidence matrix $\{t_{ij}\}$ with i as the index of N_{ex} points and j, as the index of the faces of X (not necessarily nondominated) can be constructed to facilitate the checking. (Note $t_{ij} = 1$ if the ith N_{ex}-point is contained in the jth face; otherwise, it is 0.) For a detailed development along this line, see [41].

6. APPLICATIONS AND EXTENSIONS

We shall briefly discuss four topics in this section. The first topic is an empirical study using the Value Line's ratings which gives a vivid example of multicriteria decision making in real life. The next topic will focus on methods for obtaining domination cones.

The third topic will be a classification of possible applications of domination structures and nondominated solutions. Finally we shall discuss some possible extensions of the concepts and some research problems.

6.1. An Empirical Study of the SIP

A number of books and articles have been devoted to stock market behaviors. Also a number of investment survey periodicals have been established to "help" investors. The multicriteria concept has long existed in investment analyses (see [30] for instances). However, most literature focuses on either the one dimensional comparison model or the efficient (or Pareto optimal) model. A detailed discussion describing the use of domination structures in analyzing the problem and in reaching a good investment policy can be found in [38]. Here we shall briefly describe an empirical study reported in [38]. This will motivate our later discussion.

The Value Line Investment Survey used to give weekly ratings for each stock under its survey according to four criteria (currently it gives ratings according to two criteria). These criteria were short-term performance, long-term performance, dividend income and safety (the labels of the criteria changed from time to time). The rating score of each criteria varies over {1, 2, 3, 4, 5} . Roughly speaking "1" means "best," "3" means "average," and "5" means "worst," while "2" and "4" mean "above average" and "below average" respectively. <u>Value Line</u> suggests that its reader (or potential investor) first determine a nonnegative weight for each criterion and then follow the additive weight method to select stocks in which to invest. According to Theorem 2.1, if we use strictly positive weights, the selected stocks will be nondominated with respect to the domination cone $\Lambda^{\geq}$. If we use nonnegative weights, the selected stocks will be nondominated with respect to the domination cone $\Lambda^{>} = \{d \in R^4 | d > 0\}$.

With this observation we classified each surveyed stock as an N-stock (nondominated) or a D-stock (dominated) using each of the two domination cones, $\Lambda^{>}$ and $\Lambda^{\geq}$. If the ratings and the suggestions of <u>Value Line</u> make sense and our selection of domination cones is appropriate, we would expect that the average "realized return" of the N-stocks would be better than that of the D-stocks.

We made six classifications according to the ratings on January, March and May of 1968, June and August of 1969, and January of 1970. The selection of these months was restricted more by the availability of data than by any other consideration.

Once N and D-stocks were separated, we selected a random sample from each of these groups of stocks (except that, since the N-stocks with respect to $\Lambda^{\geq}$ included only 10 to 20 stocks, we used the entire set). We then compared the average <u>realized</u> return rate of the samples from the N and D stocks for each month of the next six to thirty months (depending on the rating date; for instance, for the ratings of January 1968, we compared data for thirty months, while for ratings of January, 1970, we compared data for only six months. This was due to a limitation of available data concerning realized return). The following observations are interesting to note:

(i) The number of N-stocks with respect to $\Lambda^{\geqq}$ was very small (about 15 N-stocks out of about 1200 total stocks), while the number of N-stocks with respect to $\Lambda^{>}$ was fairly large (about 700 N-stocks out of about 1200 stocks). This is primarily due to the fact that each rating scale had only five discrete points.

(ii) Since the convex combination of N_{ex}-point is not necessarily an N-point, the sample from the N-stocks may not be a nondominated investment policy in SIP. However, the sample does test the performance of the N-stocks as a "population."

(iii) Three times out of six classifications with respect to $\Lambda^{>}$ and two times out of six classifications with respect to $\Lambda^{\geqq}$, the samples from the N-stocks had a better average realized return rate than those from the D-stocks at each month over the entire comparison period (30 months or 6 months etc. depending on the rating date). At all the other classifications the N-stocks did not have this kind of superior performance over the entire period of comparison. (In some months the sample from the D-stock group had a better average than that of the N-stock sample.) Thus we cannot conclude that the Value Line ratings and its suggestion yield useful information for the SIP. To obtain conclusive assertions we need further research. One would be interested in seeing whether or not there is a domination cone so that the N-stocks selected according to the ratings and the domination cone would always out-perform the D-stocks. We have not yet made such a study.

6.2. Methods for Finding Domination Cones

One of the main advantages of domination structures is that this method can incorporate available partial information concerning the decision maker's preferences in the decision making process. To effectively use domination structures, one immediately faces the problem of selecting domination cones. Here we suggest four possible ways to resolve this problem.

(i) A direct method: we follow the definition of domination factors and find the largest collection of such factors, we then use this collection for the domination cone.

(ii) An indirect method: we use the additive weight method or bounds on the directions of indifference among the criteria. From Theorem 2.1 and Remark 2.1, once we are given bounds for the weight vectors, we can construct a domination cone, Λ, so that the solutions from the additive weight method will be contained in Ext $[Y|\Lambda]$. Thus we can construct domination cones by knowing bounds for the weight vectors. Observe that $Y_c = \{y|\lambda\cdot y = c\}$ is orthogonal to λ. Once λ is specified, Y_c is specified as well. Note that Y_c is the isovalued plane in Y with respect to the function $\lambda\cdot y$.

Thus for any two vectors y^1, y^2 of Y_c, $y^2 - y^1$ will denote the direction of indifference with respect to the function $\lambda \cdot y$. That is, $\lambda \cdot y^1 = \lambda \cdot y^1 + \lambda \cdot (\alpha(y^2 - y^1))$ for any scalar of α. This essentially says that if the components of y are substituted according to the rate represented by $y^2 - y^1$, the value of $\lambda \cdot y$ will be unchanged. Note that Y_c has ℓ-1 dimensions. Once we can specify ℓ-1 independent vectors of the direction of indifference then we can generate Y_c. As a consequence λ which is normal to Y_c can be specified. From this point of view by specifying the bounds for the directions of indifference, we can locate corresponding bounds for the weights and then find the associated domination cone. Note that in constructing the directions of indifference, we can select one component of y and compare it one at a time with the remaining components fixed. Thus the construction of the directions of indifference is not as difficult as it may first appear to be.

(iii) A statistical method: in decision making, one may have a well defined concept such as utility, satisfaction, welfare, etc., as the major goal. However, such major goals are very difficult to measure. We may use some subgoals or criteria to measure such a major goal. For instance, in SIP, we may use the realized return in the future (or a specified period) as our major goal and use the ratings or measurement of f_1 and f_2 (for safety and expected return say) to measure the major goal. Suppose that we write the realized return r(x) of the policy x as

$$r(x) = \alpha' + \beta_1' f_1(x) + \beta_2' f_2(x) + \varepsilon \tag{6.1}$$

where α', β_1', β_2' are unknown constants; $f_1(x)$ and $f_2(x)$ are two indices as the measurements of the two subgoals for x and ε is a random variable, independently and normally distributed with mean = 0 and variance = σ^2.

Observe that (6.1) essentially is a linear multiregression model. Deleting x from the equation and using t as the index of samples, we get

$$r_t = \alpha' + \beta_1' f_{1t} + \beta_2' f_{2t} + \varepsilon_t \tag{6.2}$$

By the least square method, we can use samples to estimate α', β_1', and β_2'. Denote such estimators by a, b_1 and b_2 respectively. It can be shown that (b_1, b_2) has a bivariate normal distribution with a known covariance matrix and means equal to (β_1', β_2'). Furthermore, by preassigning a probability of confidence, say 90%, we could find an ellipse in (b_1, b_2) space so that we have 90% confidence that (β_1', β_2') will lie in the ellipse. (See Figure 4.)

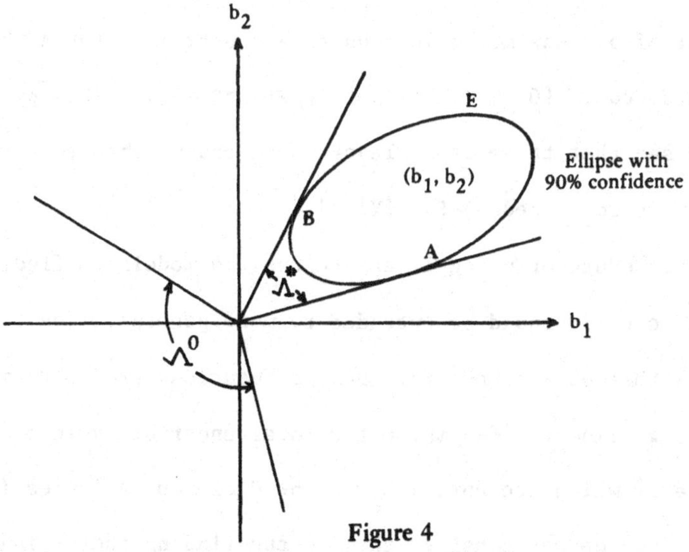

Figure 4

Observe that we have decomposed the uncertainty of the realized return r(x) into four parts. However, the unknown paramater α' and randomness of ε are independent of our decision. That is, no matter what x is, we have to bear the uncertainty of the unknown parameter α' and randomness of ε. With this understanding, we may note that to maximize the realized return r(x) under uncertainty is equivalent to maximize $\beta_1' f_1(x) + \beta_2' f_2(x)$ with a certain confidence that β_1' and β_2' will lie in a specified ellipse.

Observe that when β_1' and β_2' are fixed, $\beta_1' f_1(x) + \beta_2' f_2(x)$ is linear in f_1 and f_2. In figure 4, the ellipse E has a 90% chance of containing the true values of β_1' and β_2'. Let OA, OB be the two tangent lines to E. We see that the cone Λ^* bounded by AOB forms the minimum convex cone containing E. Since $E \subset \Lambda^*$ and Prob $\{(\beta_1', \beta_2') \in E\} = 0.9$, we see that Prob $\{(\beta_1', \beta_2') \in \Lambda^*\} \geq 0.9$. Thus the set $\cup \{Y^\circ(\lambda) | \lambda \in \Lambda^* \ \lambda \neq 0\}$ has at least a 90% chance of containing the true "optimal" solution. (Recall that $Y^\circ(\lambda)$ is the set of all maximum solutions of $\lambda \cdot y$ over Y.) Let Λ° be any convex cone contained in $\{0\} \cup \text{Int } \Lambda$ (note that $\Lambda^{**} = \bar{\Lambda}$). Then by Theorem 2.1, we see that there is at least a 90% chance that the true optimal decision is contained by Ext $[Y|\Lambda^\circ]$.

The above procedure of using linear regression models to find domination cones clearly could be extended to more general cases. We shall not pursue this extension here. Some precautions are worth mentioning. First, we linearly decompose the total uncertainty into several parts. Those parts which are unrealted to the decision variables (i.e., α and ε) although we do not consider them in the finding the domination cone, could actually affect our main goal. Next, if we use historical data (or rating) to estimate β_1 and β_2, we must check whether or not β_1 and β_2 are stable (time homogeneous). If they are not stable, certain adjustments are needed. Finally we use the regression model to estimate a domination cone which in turn is used to screen out some good alternatives, rather than to find the optimal decision. The adaptive procedure to be described later may be useful in finding the final decision. Observe that if we on one hand make ratings (f_1, f_2) for each stock, on the other hand we make decisions based on the estimated domination cone and on the ratings, then we could use the feedback concept (by checking backward) to measure the efficiency of the rating. The more stable and the larger the resulting domination cone is the better. This feedback checking is not only valuable for future decision making but also is good information for management. Later we shall discuss this concept further.

(iv) Other Methods: The successful application of a mathematical tool to a practical problem is certainly an art. It requires a thorough understanding of the problem and the available tools. The application of domination structures certainly cannot escape this requirement. Obtaining a suitable domination cone for a practical problem often depends on how well we understand the problem. For instance, in SIP, if we consider interest rates for borrowing or lending money, we might find that a great deal of inferior decisions can be eliminated and a "minimum" domination cone can consequently be constructed. In order to see this point, let us use $f_1(x)$ = -standard deviation of the return of x and $f_2(x)$ = the expected return of x as two criteria for the SIP. Now suppose we use policy x with an investment αM, $\alpha \geqq 0$. (Thus the investment distribution is $(\alpha M x_1, \ldots, \alpha M x_n)$.) If $\alpha > 1$ we have to borrow $(\alpha-1)M$ and to pay interest $(\alpha-1)RM$, where R is the interest rate for borrowing money. If $\alpha < 1$, we could put $(1-\alpha)M$ in the bank and gain interest $(1-\alpha)rM$, where r is the interest rate for "lending" money. The two criteria in terms of α and x are then given by

$$\bar{f}_1(\alpha,x) = \alpha f_1(x) \quad \text{for all} \quad \alpha \geqq 0$$

$$\bar{f}_2(\alpha,x) = \begin{cases} \alpha(f_2(x)-rM) + rM \text{ for } & \alpha \in [0,1] \\ \alpha(f_2(x)-RM) + RM \text{ for } & \alpha > 1 \end{cases}$$

(Note $\bar{f}_j(\alpha,x)$, $j = 1, 2$, are the measurements for f_j when αM is to be invested according to the policy x, and borrowing and lending money are to be executed.)

Now suppose that Y is as in Figure 5.

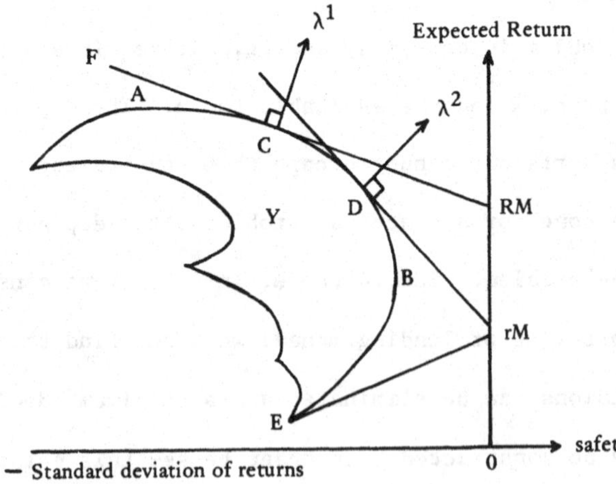

Figure 5

From $\bar{f}_j(\alpha,x)$, by introducing α, the set bounded roughly by the curve (F,C,D,rM,E) is obtainable. Now suppose that the investor is a risk-avertor (that is, when f_2 is fixed a higher f_1 is more preferred). Then in the enlarged criteria space, he would consider only those policies which yield the points on the curve (F,C,D,rM). However, these are the solutions of the additive weight method with the additive weights being bounded by the cone which is generated by λ^1 and λ^2, where λ^1 is normal to the line [C,RM] and λ^2 is normal to the line [D,rM].

According to Theorem 2.1, we then can construct its related domination cone.

6.3. Some Forms of Applications

In this section we shall briefly describe some applications of domination structures. Instead of describing the details of specific multicriteria decision problems, we shall briefly indicate the general form of several applications.

(i) Multicriteria decision making with completely known domination structures. Since the final decision must be a non-dominated solution, we can focus on the nondominated solutions. The set of optimal weights and/or the satisfying levels (described in Section 2 and 3) for each non-dominated solution can be used to help in reaching the final decision.

(ii) Multicriteria decision making with partially known domination structures. This is a very common situation in many decision problems. In Section 6.2 we described several methods for obtaining the related domination cone. The following flow chart describes an iterative procedure which can be used to reach a good decision.

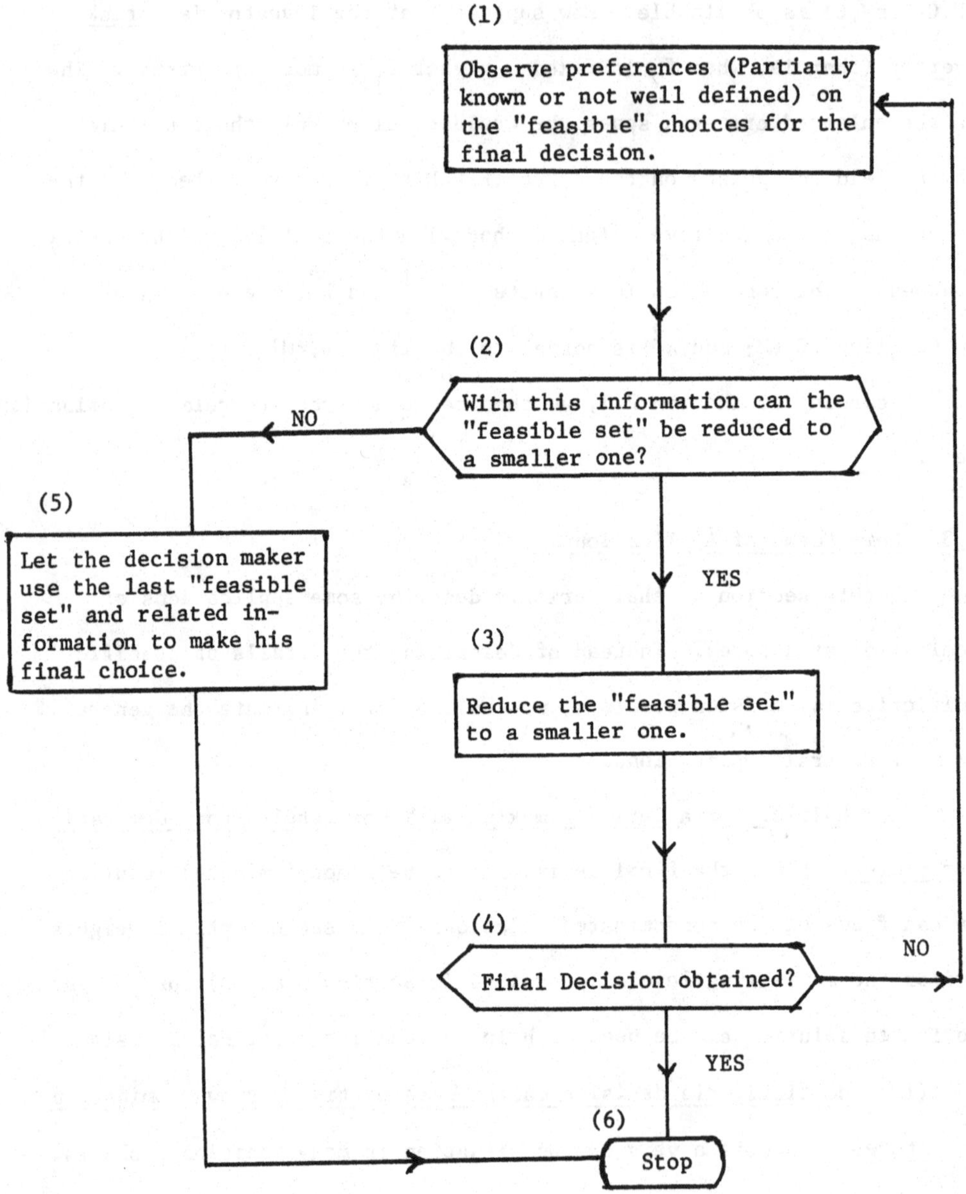
(1)
Observe preferences (Partially known or not well defined) on the "feasible" choices for the final decision.
(2)
With this information can the "feasible set" be reduced to a smaller one?
NO
YES
(3)
Reduce the "feasible set" to a smaller one.
(4)
Final Decision obtained?
NO
YES
(5)
Let the decision maker use the last "feasible set" and related information to make his final choice.
(6)
Stop

Box (1). We can use the domination structure defined on the "feasible set" as information concerning the decision-maker's preferences. The methods described in Section 6.2 will be useful for obtaining the desired domination cones.

Box (2), (3) and (5). Let Y be the feasible set and $\Lambda = \{D(y) \mid y \in Y\}$ for Box (1). Is Ext $[Y|\Lambda]$ smaller than Y? If it is, we use Ext $[Y|\Lambda]$ as our new "feasible set", go to Box (4), continue our process; otherwise, we go to Box (5) and stop. In Box (5), we could give the set of optimal weights and satisfying levels of each nondominated solution to the decision-maker. (See Section 2-3.)

Box (4) is clear. However, one may use the set of optimal weights and satisfying levels for each nondominated solution to help in reaching the final decision.

(iii) Decision Making under uncertainty. In statistical decision analysis, one usually assumes that the payoff of each decision depends on the state of nature. Suppose that the number of states of nature is finite. Denote the states by 1, 2,..., and denote the payoff of decision x at state i by $f_i(x)$. Then, clearly, we can convert the problem into our multicriteria decision model. Suppose that we have some partial information on the distribution of the states and suppose that we want to maximize the expected payoff. Then according to (ii) of Section 6.2 we can derive a related constant domination cone. Otherwise, we can begin with the domination cone, $\Lambda^{\leqq}$, and find the related nondominated solutions together with their sets of optimal weights. Note that we can normalize the weight vector (by $\sum_i \lambda_i = 1$). Then the set of normalized optimal weight vectors for each nondominated solution x can be interpreted as the set of probability distributions on the states of nature for which x yields the maximum expected payoff (with respect to the distribution).

(iv) Sensitivity Analysis and Parametric Programming. Suppose we have a single objective function which can vary throughout some given range. We may first select one representative objective function and find its maximum solution. We then study how sensitively this maximum solution responds to variations in the objective function. This is so called sensitivity analysis in mathematical programming. When the objective varies with some parameters, it is also called parametric programming. Clearly, the range of the objective function can be specified without loss of generality by a cone. We see that our domination structure analysis is very closely related to sensitivity analysis and parametric programming. Our results certainly can shed light on the area of sensitivity analysis and parametric programming. In [42], we discussed how the multicriteria simplex method can be used to perform a sensitivity analysis and to solve parametric programming problems in the linear case. We shall not repeat it here.

(v) A Feed Back Management Control Model. As indicated in Section 6.2, we can use a statistical method to estimate the domination cone for the SIP. Thus, given the ratings and the domination cone, we can make our selection from the set of the nondominated stocks. We can then compare the performance of the N-stocks and D-stocks. This will give us a feedback guidance to see if our rules for ratings and selecting domination cones are meaningful. If they are, we can continue using them or improving them.

Otherwise, a change in the rules for the ratings and selecting the domination cone must be made. This feedback process can be adopted for periodic feedback over an extended time period.

(vi) Studying the assumptions which are made by various solution concepts. We have partially discussed this topic in Section 1. A detailed treatment on a particular problem can be found in [39].

6.4. Extensions and Further Research

We have described domination structures for multicriteria decision problems. Several extensions are possible. In [2], we extended the results to domination structures in which each D(y) is a convex set rather than a convex cone. In [40], necessary and/or sufficient conditions for nondominated controls and cone convexity are reported. Once of the most important developments is to actually apply the results to practical decision problems. Those applications mentioned in Section 6.3 needed to be carried out too. K. Bergstresser and the author are working on the application of domination structures to n-person games. Some partial results have been obtained. We shall report the results when the research is completed.

7. Acknowledgement

It is a great pleasure for me to thank K. Bergstresser for his careful reading of the manuscript and some helpful suggestions. My gratitude is also extended to Mrs. M. Krause for her careful typing of this article.

REFERENCES

1. ARROW, K. J., Social Choice and Individual Values, Cowles Comission Monograph No. 12, 1951.
2. BERGSTRESSER, K., CHARNES, A. and YU, P. L., Generalization of Domination Structures and Nondominated Solutions in Multicriteria Decision Making (to appear in J. of Opt. Theory and Application).
3. BLAQUIERE, A., GERARD, G., and LEITMANN, G., Quantitative and Qualitative Games, Academic Press, New York, New York, 1969.
4. CHARNES, A. and COOPER, W. W., Management Models and Industrial Applications of Linear Programming, Vols. I and II, John Wiley and Sons, New York, New York, 1971.
5. DACUNCHA, N. O., and POLAK, E., Constrained Minimization Under Vector-Valued Criteria in Finite Dimensional Space, Journal of Mathematical Analysis and Applications, Vol. 19, pp. 103-124, 1967.
6. FERGUSON, T. S., Mathematical Statistics, A Decision Theoretic Approach, Academic Press, New York, New York, 1967.
7. FISHBURN, P. C., Utility Theory for Decision Making, John Wiley and Sons, New York, New York, 1970.
8. FREIMER, M., and YU, P. L., An Approach Toward Decision Problems with Multi-objectives, University of Rochester, Center for System Science, Report No. 72-03, 1972.
9. FREIMER, M., and YU, P. L., The Application of Compromise Solutions to Reporting Games, Appears in Game Theory As A Theory of Conflict Resolution, edited by Anatol Rapoport, D. Reidel Publishing Company, 1974.

10. FREIMER, M., and YU, P. L., Some New Results on Compromise Solutions, University of Rochester, Graduate School of Management, Series No. F7221, 1972.

11. FREIDMAN, A., Differential Games, John Wiley and Sons (Interscience Publishers), New York, New York, 1971.

12. GEOFFRION, A. M., Proper Efficiency and The Theory of Vector Maximization, Journal of Mathematical Analysis and Applications, Vol. 22, pp. 618-630, 1968.

13. GEOFFRION, A. M., DYER, J. S., and REINBERT, A., An Interactive Approach for Multicriterion Optimization with an Application to the Operation of an Academic Department, Management Science 19, No. 4, 1972, 357-368.

14. HO, Y. C., Final Report of the First International Conference on the Theory and Applications of Differential Games, Amherst, Massachusetts, 1970.

15. HOUTHAKKER, H., Revealed Preference and the Utility Function, Economica Vol. 17, 1950, pp. 159-174.

16. ISAACS, R., Differential Games, John Wiley and Sons, New York, New York, 1965.

17. ISAACS, R., Differential Games: Their Scope, Nature, and Future, Journal of Optimization Theory and Applications, Vol. 3, pp. 283-295, 1969.

18. LEITMANN, G., and Schmitendorf, W., Some Sufficient Conditions for Pareto-Optimal Control, Journal of Dynamical Systems, Measurement and Control, Vol. 95, No. 3, 1973.

19. LEITMANN, G., ROCKLIN, S., and VINCENT, T. L., A Note on Control Space Properties of Cooperative Games, Journal of Optimization Theory and Applications, Vol. 9, pp. 279-290, 1972.

20. LUCE, R. D., and RAIFFA, H., Games and Decisions, John Wiley and Sons, New York, New York, 1967.

21. MACCRIMMON, K. R., An Overview of Multiple Objective Decision Making, Appears in Multiple Criteria Decision Making, edited by J. Cochrane and M. Zeleny, University of South Carolina Press, 1973.

22. MANGASARIAN, O. L., Nonlinear Programming, McGraw-Hill, 1969.

23. RAIFFA, H., Decision Analysis, Addison-Wesley Publishing Company, Reading, Massachusetts, 1968.

24. RAIFFA, H., Preferences for Multi-Attributed Alternatives, The Rand Corporation, Memorandum No. RM-5868-POT/RC, 1969.

25. RAPOPORT, A., N-Person Game Theory--Concepts and Applications, The University of Michigan Press, Ann Arbor, Michigan, 1970.

26. ROY, B., How Outranking Relation Helps Multiple Criteria Decision Making, Appears in Multicriteria Decision Making, edited by J. L. Cochrane and M. Zeleny, University of South Carolina Press, 1973.

27. SALUKVADZE, M. E., Optimization of Vector Functionals, I, Programming of Optimal Trajectories (in Russian), Avtomatika i Telemekhanika, No. 8, pp. 5-15, 1971.

28. SALUKVADZE, M. E., Optimization of Vector Functionals, II, The Analytic Construction of Optimal Controls (in Russian), Avtomatika i Telemekhanika, No. 9, pp. 5-15, 1971.

29. SALUKVADZE, M., On the Existence of Solutions in Problems of Optimization Under Vector-Valued Criteria, Journal of Optimization Theory and Applications, Vol. 13, No. 2, 1974.

30. SHARPE, W., Portfolio Theory and Capital Markets, McGraw-Hill, 1970.

31. STADLER, W., Preference Optimality (also his lecture notes for this seminar).

32. STALFORD, H. L., Criteria for Pareto-Optimality in Cooperative Differential Games, Journal of Optimization Theory and Applications, Vol. 9, pp. 391-398, 1972.

33. VINCENT, T. L., and LEITMANN, G., Control-Space Properties of Cooperative Games, Journal of Optimization Theory and Applications, Vol. 6, pp. 91-113, 1970.

34. VON NEUMANN, J., and MORGENSTERN, O., Theory of Games and Economic Behavior, Princeton University Press, Princeton, New Jersey, 1947.

35. YU, P. L., A Class of Solutions for Group Decision Problems, Management Science, Vol. 19, No. 8, 1973.

36. YU, P. L., Cone Convexity, Cone Extreme Points and Nondominated Solutions in Decision Problems with multiobjectives, Journal of Optimization Theory and Applications, Vol. 14, No. 3, Sept., 1974.

37. YU, P. L., Introduction to Domination Structures in Multicriteria Decision Problems, Appears in Multiple Criteria Decision Making edited by J. Cochrane and M. Zeleny, University of South Carolina Press, 1973.

38. YU, P. L., Nondominated Investment Policies in Stock Markets (Including an Empirical Study), Systems Analysis Program F7222,

University of Rochester, 1973.

39. YU, P. L. and LEITMANN, G., Compromise Solutions, Domination Structures and Salukvadze's Solutions, Journal of Optimization Theory and Application, Vol. 13, No. 3, March, 1974.

40. YU, P. L. and LEITMANN, G., Nondominated Decisions and Cone Convexity in Dynamic Multicriteria Decision Problems, Journal of Optimization Theory and Applications, Vol. 14, No. 5, November, 1974.

41. YU, P. L. and ZELENY, M., The Set of all Nondominated Solutions in the Linear Cases and a Multicriteria Simplex Method, Journal of Mathematical Analysis and Applications, Vol. 49, No. 2, Feb.]975.

42. YU, P. L. and ZELENY, M., On Some Linear Multi-Parametric Programs, CSS 73-05, Center for System Science, University of Rochester, Rochester, New York, 1973.

43. ZADEH, L. A., Optimality and Non-Scalar-Valued Performance Criteria, IEEE Transactions on Automatic Control, Vol. AC-8, pp. 59-60, 1963.

44. ZELENY, M., Linear Multiobjective Programming, The University of Rochester, Graduate School of Management, Ph.D. Thesis, 1972.

ON SOME BROAD CLASSES OF VECTOR OPTIMAL DECISIONS AND THEIR CHARACTERIZATION

A. MARZOLLO W. UKOVICH

Electrical Engineering Department, University of Trieste
and
International Centre for Mechanical Sciences, Udine

1.- INTRODUCTION.

We shall consider situations in which some " decision maker" has to choose, in a set of feasible decisions, a decision which may be considered as the " best" according to some finite set of criteria.

To be more precise, and using for vector comparisons the same notations as in Stadler (p.128 in this volume), the following will be the " data" of our problem:

- a " feasible set" X in R^n, which we shall usually suppose to be given by:

$$X = \{ x \mid g(x) \geq 0 \}$$

where g maps R^n into R^k,

- an m-vector valued objective function $f = (f_1, f_2 \dots f_m)$, (with $m > 1$) defined on some open set containing X and whose components we try to maximize.

Notice that we don't make for the moment any " a priori" assumption on the way in which the best is defined out of the objective function. As it will be seen in the sequel, this also implies that no further hypothesis is made on the real structure of the decision maker: it may be a single person considering different aspects (or " attributes") of the consequences of his decisions, or even a group. if agents, each one of them contributing to such a choice(in a cooperative, or noncooperative, or indipendent way, or in any other mixed form) according to one (or more) objective (or " utility", in such a case) function.

Since there are more than one objective functions, but there is no way of getting from them an unambiguous definition of " best", our original problem may be seen as resulting from the union of two different subproblems: the definition of the best and the search for it. We shall usually refer to the first phase as the " aggregating" one, and to the second as the "maximizing" one.

In Part I of the present contribution we shall critically review some general concepts and solutions concerning the aggregation problem, and in Part II we shall give some new results for the maximum problem in very wide framework.

In Part I we shall put special enphasis on the solutions for the aggregation problem proposed by authors mainly dealing with the maximum problem. In fact, as we shall see, the results for the latter problem strongly depend on what solution has been accepted for the former one.

Roughly speaking, the literature dealing with situation similar to ours may be divided into two main groups, depending on whither the aggregation or the maximum apsect are mainly stressed. Our interest in scanning the literature is somehow intermediate between the ones of such two groups, since we look for the relations between them. This is why we shall skip out the whole wide body of the so-called "decision making theory". In fact, its main concern consists in proposing some meaningful aggregation procedure that is also acceptable from a computational point of view, or in suggesting some "iterative" method leading to the best solution by successive approximations of what is best, where at each step both the aggregation and the maximization phases are used. In the former case no relation between aggregation and maximization is explored at all, whereas in the latter such a connection is too strict to show its meaning in a fully general way. (For an exhaustive introduction to the decision making theory literature, see for example Roy 1971, Mac Grimmon 1973, Keeney and Raiffa 1975).

Furthermore, it must be stressed that we shall not give an exhaustive review of the many definitions of "best" that have been proposed (for such a review, see for example Yu, this volume) but we are interested only in making some considerations about a narrow set of definitions, which are the most relevant ones from the point of view of our general formulation.

The original aspects of the present contributions are contained only in Part II. We use there from the beginning the concept of optimal decisions as the ones "non dominated" with respect to a cone, according to Yu, and moreover we clearly distinguish among (globally) "weakly", "ordinarily" and "strictly" non dominated decisions x^o. For each kind of optimality we define the corresponding "local" version; and the "differential" one, obtained by substituting to f and g their linear approximations at x^o.

Since, as it is intuitive, from a numerical point of view the conditions for differential optimalities are the easiest to be checked, the practical interest is clear of finding sufficient conditions on f and g for a differentially optimal decision x^o (of any kind) to be also locally (and possibly globally) optimal. The search is then for the maximal relaxation of such conditions, which is carried out in Part III,3 and in Part II,4, using two different approaches. In Part II,3 the weakest conditions are also given, which permit to infer each one of the above defined optimalities from each other one.

PART I

I.1.- GENERAL CONSIDERATIONS ON THE CONCEPT OF PARETO OPTIMALITY.

The most usual definition of " best" is the definition of the so-called " Pareto optimality": a decision x^o is Pareto-optimal if no feasible x' exists such that $f(x') > f(x^o)$ (Later on, this type of Pareto Optimality will be refered to as " ordinary" Pareto optimality).

The above definition owes its name to the economist Wilfredo Pareto, who used it (Pareto 1896) in the general fromework of his economic theory. Since then, such a definition has been widely used and investigated, not only in the economic area (see Koopman 1951, Karlin 1959, Debreu 1959, etc.), but also as a quite general concept of solution for mathematical programming problems (see Kuhn and Tucker 1951, Geoffrion 1966, 1967 a, 1967 b, Chu 1970, Smale 1973, 1974, Wan 1975), game theoretic and control problems, both in the static (see for example Luce and Raiffa 1957, Zadeh 1963, De Cunha and Polak 1967, Schmitendorf 1973) and in the dinamic case (see for example Chang 1966, Storr and Ho 1969 a, 1969 b, Vincent and Leitmann 1970, Rekasius and Schmitendorf 1971, Leitmann, Rocklin and Vincent 1972, Blaquière, Juricek and Wiese 1972, Haurie 1973, Leitmann and Schimetendorf

1973, Leitmann 1974, Haurie and Delfour 1974, Leitmann this volume, Blaquière this volume).

It is interesting to define the Pareto optimal decisions by first introducing the binary " Pareto preference-or-indefference relation" $\bar{R}$, defined on the cartesian product $f(X) \times f(X)$ as follows:

$$(y^1, y^2) \in \bar{R} \iff y^1 \quad y^2$$

and then characterizing the Pareto optimal decisions as the ones mapped by f into its maximal set.

According to the current terminology (see for example Sen 1970, Birkhoff 1948), y^o belongs to the maximal set of $\bar{R}$ of no y' in the domain of $\bar{R}$ exists such that (y', y^o) is in $\bar{R}$, and (y^o, y') is not. (In such a more general fromework, the relation $\bar{R}$ may be seen as the result of the aggregation phase, implicity defining the best as the counterimage, by f, of its maximal set).

In an equivalent way, defining the " binary Pareto preference" relation $\bar{P}$, by

$$(y^1,y^2)\epsilon\bar{P} \iff (y^1,y^2)\epsilon\bar{R},(y^2,y^1)\not\epsilon\bar{R}$$

y^o belongs to the maximal set of $\bar{R}$ iff no y' is in the domain of $\bar{R}$ such that $(y',y^o)\epsilon\bar{P}$, (that is, is prefered to y^o).

The definition of " best" as Pareto optimal is somehow the most general one for our problem as expressed in the introduction in the sense that it may be considered as the smallest requirement one can put on a " best" decision if no information is provided about the " preference structure" in our problem, except the objective function. In fact, each component of f is considered as having the same importance in determining whether a given decision is optimal or not.

For a deeper understanding of the above fundamental statement let us refer to each component of the objective function as the utility function of a distinct " agent" (we use such a " personalizing" interpretation just for the sake of simplicity in exploring the main features of Pareto Optimality; all conclusions, however, still hold in a general formulation). Then the above definition of Pareto preference-or-indefference relation may be seen as implying same kind of cooperation among the agents, since according to it a decision x^o is preferred to another one x' if the utility function of no agent decreases passing from x^o to x', and at least one of them increases. In other words, each agent is willing to give up any of his possible marginal gains of these could be achieved only at the expenses of someone else. Consider now the following method to test the optimality of a given feasible decision x^o: each agent (say the i-th) indicates two sets: the set $C^i_{x^o}$ of feasible decisions which are " preferred or indifferent " to x^o with respect to his personal utility function, that is

$$C^i_{x^o} = \left\{ x \mid x \in X,\ f_i(x) \geq f_i(x^o) \right\}$$

and the set $I^i_{x^o}$ of feasible decisions which are " indifferent" to x^o, that is

$$I^i_{x^o} = \left\{ x \mid x \in X\ ,\ f_i(x) = f_i(x^o) \right\} .$$

Then x^o is Pareto optimal if

$$\bigcap_i C^i_{x^o} = \bigcap_i I^i_{x^o} \qquad \text{I,1(1)}$$

(where the intersections are extended to all indices i=1...m). Observe that $x^o \varepsilon I^i_{x^o}$, for all i. Furthermore, $\bigcap_i C^i_{x^o} \supset \bigcap_i I^i_{x^o}$, since $C^i_{x^o} \supset I^i_{x^o}$, for all i. The above consideration shows that, in all situations, each agent has the right to declare his desires and to require that they contribute to determine the optimality of x^o, whatever they are. In this sense, we may say that each agent has equal rights, whatever is the set of the decisions preferable to x^o for him. Clearly, such an " equality" among the agents does not exclude that a " vetoing" situation may happen, for example if $C^i_{x^o}$ reduces to x^o for some agent: in fact, in such a case the Pareto optimality of x^o is guaranteed (since both sides of equation I.1,(1) reduce to x^o), without any further consideration for the other agents preferences. The above " equality" must be therefore intended in the restricted sense that no $C^i_{x^o}$ (nor $I^i_{x^o}$) may be disregarded " a priori", that is because of its shape. The possibility for a " vetoing situation" may be then seen as an unavoidable consequence of our " equality principle", which implies the conservative principle of requiring unanimity for deciding that decision is " better" than another one, thus allowing any single agent to let a decision become Pareto optimal.

An interesting consequence of the equality principle as stated above is the so called no intercomparability among the components of the objective function, which means that the relative amplitudes of the possible marginal gains and losses of the agents do not decide about the Pareto optimality of a given decision. In fact the equality principle implies that also the shapes of the images of the sets $C^i_{x^o}$ on the reals through f_k for every pair i, k are not a priori decisive about the Pareto optimality of given decision x^o. Indeed $f_k(C^i_{x^o}) - f_k(x^o)$ is the set of possible marginal gains for the i-th player if i = k, whereas it represents the set of marginal gains (or losses), of the k-th agent if he would accept to depart from the decision x^o according to the preferences of the i-th agent, if i ≠ k. Clearly, whereas the equality principle implies non-intercomparability, the viceversa is in general not true, since the equality requires that no property of $C^i_{x^o}$ has any a priori relevance for deciding about optimality, whereas the no intercomparability deals only with images through f_k. An other fundamental property of Pareto optimality may be seen as a consequence of the equality principle: its invariance under any monotonic transformation of the objective function. That is, if x^o is a Pareto optimal decision with respect to a given objective function f, it is also Pareto optimal with respect to f', if f' is obtained from f by monotonically transforming each of its components; moreover, such a transformation need not be the same for all components. In fact, neither $C^i_{x^o}$ nor $I^i_{x^o}$ are modified by such a transformation. The above property is often referred to by stating that Pareto optimality has a " purely ordinal" character, since it is influenced only by the ordering which the f_i' s induce on the set of the feasible decisions.

I.2.- DISTINCTIONS IN PARETO OPTIMALITY: IMPROPERNESS

The search for alternative definitions of what is " best" is motivated by some drawbacks suffered by the concept of Pareto optimality. All of them may be seen as consequences of its generality, which we characterized by the fact that it obeys to the equality principle, and therefore no intercomparison holds.

To use a someway suggestive comparison, if no intercomparison is accepted, the situation is similar to the one in wich, in the presence of a given number of different possible events, one feels temped to ascribe an equal probability to each of them. Such an attitude, even if questionable, may be seen in practice at least as good as any other one if no further insight may be obtained about the meaning and the structure of the definite problem we are dealing with.

Other, more practical, considerations stem the fact that the effective usefulness of determining the set of all Pareto optimal points (or, simply, the Pareto set) is often quite short to give any valuable suggestion on what decision has to be ultimately taken. In fact, us we shall see, there is no way of finding any convincing criterion to compare each other two Pareto optimal points without introducing among the components of the objective function some aggregation rule which must be sharper than the simple unanimity rule, hence implying some intercomparison. Therefore, unless the Pareto set is a singleton, and if intercomparison is not allowed, no reason may be derived from the given criteria to select the particular decision to be implemented.

To justify the above statement, suppose that the Pareto set of a given multicriteria optimization problem is taken as the feasible set for a " second level" scalar criterion optimization problem, as some authors suggest (see Zadeh 1963). Clearly the second level criterion cannot have the same relevance as the first level criteria, since otherwise it would be used with them in the first problem. By the same argoment, the second level criterion must somehow depend on the shape of the Pareto set of the first problem, hence on the first level criteria. Suppose now, for the sake of simplicity, that the second level criterion is represented by the scalar-valued function f_o : we proved that it must " depend" on the f_i' s. That is, there must be a map f'_o such that $f_o = f'_o\,(f_1,f_2,\dots f_m)$. Clearly, f'_o implicity represents the way in which the first-level objective functions f_i are aggregated in order to give f_o. The conclusion may then be drawn that some intercomparison of the f_i ' s is implied by taking the Pareto set as the feasible set for a second level optimization problem.

The last result is a fundamental one to fully understand the rest of the present section in which some distinctions are presented that have been proposed for the definition of Pareto optimality.

A first example is the distinction, introduced by Kuhn and Tucker in their pioneering paper (Kuhn and Tucker 1950), between "proper" and " improper" Pareto optimal points.

Following the definition they give, a Pareto optimal point x^o is proper if there is no feasible direction in which the (first-order) directional derivatives of the objective functions are neither nonnegative nor simultaneously equal to zero. That excluding improper Pareto optimal points implies objective inter-comparison is explicity admitted by the authors, since they propose the argument (further developed by Klinger in (Klinger 1964) and (Klinger 1967)) that improper points enjoy the " undesirable property" that there is a direction in which a possible first-order increase of (at least) one component of the objective function corresponds to a higher order decrease of some other. Clearly, the preference comparison introduced by Kuhn and Tucker might be considered of only minor relevance, since it does not concerny the numerical values of the objective functions, but only the magnitude order of their rate of increase. In fact, improperness is invariant under (positive) linear transformation of the objective functions. On the other hand, however, a fundamental remark should be made about the concept of properness: only first-order derivatives are considered. This is the reason why nonproper Pareto optimal points always concern a first-order marginal gain against higher-order marginal losses, whereas it seems to be equally " undesirable" that higer-order infinitesimal marginal losses may in general prevent lower-(not necessarily first) order infinitesimal marginal gains. Following the definition of Kuhn and Tucker, the last situation produces either a nonproper or a proper Pareto optimal point depending on whether the possible gain is respectively of the first-order or not. It might be obvious that in most cases such a discrimination makes no sense, since not only the magnitude order of the objective functions rates of increase are compared, but also the effective value of one of them must be considered. This makes improperness not invariant not only if a strictly monotone transformation is applied to any component of objective function (there is no hope for this, since preference intercomparison is necessarily implied) but also if some strictly monotone transformation is applied to all components of the objective functions.

To overcome this drawback, Geoffrion proposed, in Geoffrion 1967b, a new definition, in which a proper point is characterized by the property that the ratio of the possible marginal gain and the possible marginal loss with respect to any two elements of the objective function is bounded. To be more precise, x^o is a proper Pareto optimal point in Geoffrion's sense if there exists a scalar $M > 0$ such that, for each i, the relation

$$\frac{f_i(x) - f_i(x^o)}{f_j(x^o) - f_j(x)} \leqslant M \qquad \text{I.2,(1)}$$

holds for some j such that $f_j(x) < f_j(x^o)$, whenever $x \in X$ and such that $f_i(x) > f_i(x^o)$.

The " global" character of Geoffrion's definition must be

stressed, because it produces much more non proper points than one might expect by considering it as an obvious generalization of the " undesirable property" . The reason is that to the second formulation of the " undesirable property" it was implicity given only a " local" validity, in the sense that only infinitesimal variations of the objective functions were considered. (Obviously, the term " local" is used here with respect to the objective space $R^m \supset f(X)$).

Nevertheless, it is reasonable to extend the " undesirable property" to have a global validity by considering also arbitrarily large variations of the objective functions. It may therefore be formulated as follows: it is desirable that for any Pareto optimal point there does not exist another feasible point giving a marginal gain for an element of the objective function and not giving a marginal loss of the same magnitude order for some other one. Then it is clear that improper Pareto optimality in the Geoffrion's formulation exactly matches with global undesirability.

If may be immediately observed that global undesirability is invariant under simultaneous strictly monotone transformations.

As usual, objective functions concavity and feasible set convexity guarantee that local properties become global ones. For nonproperness, this results from the fact that no concave function may tend to $+\infty$ in a finite point, whereas if it has a finite limit along any direction, it must be nondecreasing along that direction. Hence there cannot exist a Pareto optimal point with marginal gains unbounded above whereas all marginal losses are bounded below. Remembering that nonproperness is invariant under simultaneous strictly monotone transformation, we conclude that in the case of concave objective functions the properness condition cannot be violated because of the numerator unboundedness of the left-hand side of I.2,(1).

Before concluding our presentation of properness for Pareto optimal points, some considerations are appropriate concerning the validity of such a concept. First of all, it may be pointed out that whereas it is considerably hard formally characterize (and actually compute) the set of all Pareto optimal points in the most general case, all difficulties are drastically reduced, if the problem is restricted to proper Pareto optimal points (see Marzollo and Ukovich 1974, 1975 versus Geoffrion 1967 b). Such an useful simplification may be obtained if the principle is accepted that non proper Pareto optimal points are not to be considered as giving satisfactory decisions, as it is for the non-optimal ones. However, such an assumption does not seem to have general validity.

Consider for example a point $\bar{x}$ in which the maximum is achieved of the i-th component of the objective function. Obviously, $\bar{x}$ is Pareto optimal; however, supposing that f_i is derivable in it, its local properness does not longer depend on f_i, unless some very special conditions are verified.

Such an anomaly is produced by the fact that in $\bar{x}$ only agents' tendencies to move to increase their gains are considered in order to get properness, whereas i-th agent's aspiration not to move is absolutely disregarded.

Again we see that properness does not consider all players agents' preferences with the some equity, thereby violating the equality principle. Clearly, most "vetoing situations" are eliminated in such a way, but consider again the case in which the components of f are the utility functions for different agents: even if we suppose that they act in a cooperative way, it might be hand to persuade the i-th agent to give up sticking objective!

I.3.- MORE GENERAL DEFINITIONS OF OPTIMALITY: NONDOMINATION.

Once the principle is accepted that criteria intercomparison may be admitted at some extent, much broader generalizations of the concept of Pareto optimality may be presented, resulting for example in the definition of nondomination with respect to a cone, as it has been proposed by Yu (see Yu, this volume, for an exaustive discussion on this concept and of its properties).

According to Yu's definition, a decision $x^o \varepsilon X$ is said to be nondominated with respect to a given domination cone D if no $x' \varepsilon X$ exists such that $f(x') \neq f(x^o)$, and $f(x') - f(x^o) \varepsilon - D$.

We first consider the case in which $D \supset N = \{ y, y \varepsilon R^m, y \leq 0 \}$, $D \neq N$. Such a condition means that a Pareto optimal decision x^o is no longer satisfactory if there exists another one x' allowing some marginal gains for some agent at the expenses of other agent's marginal losses, the ratios between such gains end losses being such that $f(x') - f(x^o) \varepsilon - D$. From a practical point of view, the Pareto optimal points are now divided into two subsets A and B such that it is worth while to suffer a (minor) marginal loss with respect to some component of the objective function in trading a point in A with some point in B in order to get a (greater) marginal gain with respect to some other component.

Some earlier ideas about cone domination may be found in a paper by Zadeh (Zadeh 1963). It is interesting to mention them here because, even concerning a quite restricted problem, they do not involve any critera intercomparison. Moreover, they point out an useful relation between Pareto optimality and decisions domination when all objective functions are linear. In fact, the problem of finding the Pareto optimal decisions of a given feasible set X with respect to a given finite set of noncomparable linear homogegeous objective functions $f_i(x) = \langle c_i , x \rangle$, i = 1...m may be handled and solved in the decision space as the problem of finding the points x^o of X such that no $x' \neq x^o$ in X exists such that $x' \in x^o - D$ where $-D$ is the polyhedral convex cone generated by the vectors c_i. In this case, the domination cone D lies in R^n and fully replaces the role of the objective function, thereby making it superfluous to take into consideration the objective space f(X). (The above sketched approach may be obviously generalized to the case of infinitely many objective functions, resulting in a nonpolyhedral domination cone).

As it has been said, intercomparison is heavily introduced in the concept of nondominated solutions with respect to domination cones containing N, since some trade-off is admitted between marginal losses and gains. Moreover, it is clear that the generalization that nondomination introduces on Pareto optimality follows the same line as properness did, since infinite and infinitesimal orders of marginal gain and losses were compared there, whereas their finite values are considered here. Nevertheless, it is surprising to see that in general the set of all proper Pareto optimal points cannot be obtained as the set of nondominated points with respect to any domination cone. A formal proof of such a property is not trivial, and is omitted here for the sake of brevity. The logical independence between the concepts of domination and properness then suggests to distinguish between properly and nonproperly nondominated decisions.

Some special situations, depending on the properties of the domination cone, deserve now some attention: consider first the case of D containing a full-dimension halfspace, but no proper subspaces: that is, consisting of the union of the origin and of on open halfspace whose boundary is a full dimension subspace. Then we have full objectives intercomparability: there exists a well defined trade-off ratio between any two components of the objective function, which allows us to express all such components in a single unity (for example, money), thereby reducing the problem to the search for decisions maximizing their sum, weighed with the components of any(outward) normal to such domination cone.

From this point of view, the general case in which the domination cone does not contain any affine set, but contains P, may be considered as resulting from a situation of " imperfect", or " partial", intercomparability, in which some trade-offs between the criteria are admitted, but their ratios are not fixed: they are only bounded to lie within some interval. (See also Yu 1967, example 1.3 and remark 1.4). Therefore, in the other extreme case, that is in Pareto optimality, trade-offs between criteria are inconsistent, since their ratios are not bounded in any way. Consider now the case of a domination cone containing some nontrivial subspace: this implies that it contains two " domination directions" that are opposite, and it is easy to see that the strange situation may then be produced in which two decisions dominate each other, possibly preventing each other, in this way, from being optimal. As a consequence, the set of the " optimal objectives" (that is, the points $y^o = f(x^o)$, where x^o is a non-dominated decision) either is empty or consists of a singleton. Indeed, since two optimal objectives, say y' and y", cannot dominate each other, neither of the vectors y' - y" and y" - y' may lie in the domination cone, which is impossible.

This argument suggests that if the domination cone contains some proper subspace, two kinds of dominated decisions must be distinguished: the ones dominating all decisions by which they are dominated, and the ones for which there is at least a decision dominating them, but that they do not dominate. Roughly speaking, the dominated decisions of the first kind above might be expected to be somehow " less dominated" than the other ones. (Remark that such a ranking of dominated decisions does not involve any criteria intercomparison at all).

Further insight about the above situation may be gained by adopting the concepts and terminology of the binary preference relations, as we did for Pareto optimality. According to the definition of Yu, define the binary domination preference-or-indifference relation R on the cartesian product $f(X) \times f(X)$ as follows:

$$(y^1, y^2) \in R \iff y^2 \in y^1 + D. \qquad I,3,(1)$$

Remarke that R is reflexive for D closed, and it is antisymmetric (that is, $(y^1, y^2) \in R$ and $(y^2, y^1) \in R$ imply $y^1 = y^2$ for all y^1, y^2) if D contains no nontrivial subspace. In such a situation, the nondominated decisions exactly coincide again with the ones mapped by f into the maximal set of R. In fact, the binary " domination preference" relation P, which is again defined by

$$(y^1, y^2) \in P \iff (y^1, y^2) \in R, (y^2, y^1) \notin R ,$$

coincides with R whenever $y^1 \neq y^2$. However, if D contains a nontrivial subspace, the above coincidence does not longer hold in general.

For including in the definition of nondominated decisions the case of domination cones containing a nontrivial subspace it is then possible to adopt a more general definition of nondominancy (see Marzollo, Ukovich 1975), which formally agrees with the formulation of Pareto optimality we expressed using the binary relation $\bar{R}$: a decision $x^o \in X$ is nondominated with respect to a given cone D if it is mapped by f into the maximal set of R, as defined in I,3,(1).

We conclude this paragraph with considerations on the case in which the domination cone is properly contained in N. Suppose for example that D = int N (see for example Aubin 1971); then for a decision to be dominated it is required that another one exists giving some (positive) marginal gain with respect to each component of f. Such a condition may be seen as requiring a more strict cooperation among all agents than in the case of the Pareto optimality: they modify their decision only if each one is motivated to do so. In this way some kind of " inertia" is added to the agents' behaviour, since not only a marginal loss for some of them, but also indifference may prevent marginal gains of the other ones.

In the next sections, we shall call all the nondominated points with respect to int N " weakly Pareto optimal".

Similar inertial effects are always present whenever the domination cone is properly contained in N. In fact, the corresponding set of the nondominated decisions contains the Pareto optimal set, and this means that there are some decisions which the players think it is not worth while to change, even if some marginal gain could be realized without suffering any marginal loss. Such " worth while" considerations seem to be peculiar for this class of problems, in the same way as criteria intercomparison is for domination with respect to cones properly containing N.

PART II

As mentioned in the Introduction, whereas Part I has been only a critical review of some basic concepts, Part II contains the results of the present contribution. Namely, in Paragraph II,3 some theorems will be proven which guarantee the equivalence between the different kinds of "local" and "differential" optimality (or non-dominacy) which are defined in Paragraph II,1; and in Paragraph II,4 some broad and different sufficient conditions will be given which guarantee differentially optimal decisions to be also locally optimal. As it will be further remarked in the conclusions, these conditions have a special importance, since both they and differential optimalities have a practical test for being checked.

II.1.- DEFINITIONS AND REMARKS

We devote the first paragraph of this part to the definition of the nine kinds of optimality we shall refer to, and to some interesting comments and remarks.

We consider again the set X of feasible decisions given by

$$X = \{ x, \mid g(x) \geq 0 \}$$

where g maps R^n into R^k, and the objective function $f = f_1, f_2 \dots f_m$, with $m > 1$, is defined on some open set containing X.

We consider three kinds of orderings relative to a given closed and convex cone $C \subset R^m$. They are obtained from the binary relation R, defined on $f(X) \times f(X)$ by

$$(y^1, y^2) \in R \iff y^1 \in y^2 + V \qquad \text{II,1(1)}$$

when we take the set V:

i) equal to C, or

ii) equal to $C' = C/0$, or

iii) equal to int C (which we always suppose non empty).

In correspondence, such relations will be denoted by $R^{\geqslant}$, $R^{>}$, $R^{\gg}$. For the sake of simplicity, we shall often use the notation $y^1 \geqslant' y^2$ for $(y^1, y^2) \in R$, $y^1 >' y^2$ for $(y^1, y^2) \in R$, $y^1 >>' y^2$ for $(y^1, y^2) \in R$.

We thus obtain three kinds of global optimality by substituting the relation $R^{\geqslant}$, $R^{>}$ or $R^{>>}$ to R in the following definition:

$x^o \in X$ is (globally) optimal if no $x^1 \neq x^o$ exists in X such that $(f(x^1), f(x^o)) \in R$.

Remark that, when considering $R^{>}$ and $R^{>>}$, the condition $x^1 \neq x^o$ may be dropped. For the sake of clarity we explicitely state the definitions:

<u>Definition II,1,1</u>

$x^o \in X$ is a "globally strictly optimal" (G.S.O.) decision if no $x^1 \neq x^o$ exists in X such that

$$f(x^1) - f(x^o) \in C,$$

that is $f(x^1) \geqslant' f(x^o)$;

<u>Definition II,1,2</u>

$x^o \in X$ is a "globally ordinarily optimal" (G.O.O.) decision if no

x^1 exists in X such that

$$f(x^1) - f(x^o) \in C^1,$$

that is, $f(x^1) >' f(x^o)$;

Definition II,1,3

$x^o \in X$ is a "ordinarily weakly optimal" (O.W.O.) decision if no x^1 exists in X such that

$$f(x^1) - f(x^o) \in \text{int } C,$$

that is, $f(x^1) \gg' f(x^o)$.

The terminology we use is justified by the above definitions. In fact, strict optimality implies the ordinary one, which in turn implies weak optimality.

Some remarks concerning the above concepts are appropriate:

i) The relations $R^{>}$ and $R^{\gg}$ appearing in definitions II,2 and II,3, are not reflexive, since $0 \notin C^1$ and $0 \notin \text{int } C$, whereas $R^{\geq}$ is. That is why we did not use in our definitions the concept of "maximality", which is based on the binary relation P

$$(y^1, y^2) \in P \Longleftrightarrow (y^1, y^2) \in R, (y^2, y^1) \notin R,$$

and therefore the maximal sets of $R^{\geq}$ and $R^{>}$ coincide, as it may be easily checked, a circumstance which does not occur following our definitions.

ii) These definitions of optimality are actually a specification of the definition of "non dominacy" given by Yu (see for example Yu, this volume), in the sense that a clear distinction is made a priori between the case $V = C$ (which coincides with Yu's non-domination with respect to the domination cone $D = - C$), $V = C^1 = C/0$, $V = \text{int } C$.

iii) In the particular case of $C = N^+$, the non-negative orthant of R^m,

$$N^+ = \{ y \mid y \in R^m, \quad y \geq 0 \}$$

global ordinary optimality according to definition II,1,2 reduces to the standard Pareto-optimality, whereas strict optimality of definition II,1,1 is sometimes mentioned in the literature in its local version (see further in this paper) as "strict Pareto optimality" (see Smale 1973,1974, Wan 1975), and weak optimality of definition II,3 also becomes a kind of Pareto-optimality, which we could call " weak Pareto optimality", whose meaning was discussed at the end of paragraph I,3.

Remembering the considerations about "equality" of (ordinary) Pareto optimality we exposed in paragraph I.1, we give similar interpretations of the two other kinds of optimality. For weak Pareto optimality, each agent indicates the set B_i of the feasible decisions which are (strongly) better than x^o with respect to his personal criterion: then x^o is weakly Pareto optimal if, and only if, the intersection of all such B_i's is empty. For strict Pareto optimality, each agent indicates the set C_i of feasible decisions which are "better - or - equal" for him than x^o: it is strictly optimal if, and only if, the intersection of all such C_i's contains only x^o. Then the conclusion may be drawn that Pareto optimality, of any kind, always satisfies the " equality principle" quoted above.

iv) Remember also that, as it is well known, if C is a polyhedrical closed cone, the ordinary optimality of definition II,2 coincides with ordinary Pareto optimality with respect to the new objective function Hf, where the rows of the matrix H are the generators of the (polyedrical closed) cone $- C^o$, where $C^o \in R^m$ is the "dual" to C:

$$C^o = \left\{ z \mid z, c \leq 0, c \in C \right\}$$

v) Strict optimality is equivalent to ordinary optimality, if f is injective.

The three kinds of optimality, as defined above, are seldom

easy to characterise, unless some special conditions (e.g. convexity of $f(X)$) are verified: we therefore consider their "local" version, instead of the "global" one we formulated, by limiting the search for x^1 to some neighbourhood of x^o. We thus obtain the following definitions:

Definition II,1,1',2',3'

$x^o \in X$ is a "locally (strictly, L.S.O.) (ordinarily, L.O.O.) (weakly, L.W.O.) decision iff a neighbourhood I of x^o exists such that for no $x^1 \in I \cap X$

$$f(x^1) - f(x^o) \in (C)\ (C^1)\ (\text{int } C)$$

that is

$$f(x^1)\ (\geqslant')\ (>')\ (>>')\ f(x^o).$$

We shall often use an alternative formulation of the previous definitions of local optimality, which is clearly equivalent, as it may easily be checked:

$x^o \in X$ is locally (strictly) (ordinarily) (weakly) optimal if no infinite sequence $\{x^k\}$ exists such that $x^k \neq 0$, $x^k \to 0$, and, for all k, $f(x^o + x^k) - f(x^o) \in (C)\ (C^1)\ (\text{int } C)$.

Global optimality of x^o obviously implies local optimality of the same kind.

Under some concavity-type hypotheses, which it is trivial to determine, the viceversa is also true. For example, taking for simplicity $C = N^+$, and referring to Mangasarian 1969, ch. 9 for terminology, it is easy to see that if X is convex (and this is true if the components g_i, $i = 1 \ldots k$, of g are quasi-concave),

i) local strict optimality of x^o implies global strict optimality of x^o if the components f_i, $i = 1 \ldots m$, of f are quasi-concave in x^o.

ii) local ordinary optimality of x^o implies global ordinary optimality of x^o if $f(x) \geqslant f(x^o)$, implies

$$f\left[(1-\lambda)x+\lambda x^{o}\right] \geqslant f(x^{o}),\ \forall\, x \in X,\ 0<\lambda<1.$$

The condition is an obvious extension to vectors of quasi-concavity of scalar functions.

iii) local weak optimality of x^o implies global weak optimality of x^o if the f_i's are strictly quasi-concave.

Much more interesting will be to find conditions which permit to characterize the local optimalities we defined. This will be done in paragraphes II,3 and II,4, where we shall use the definitions we now give for the "differentail" version of the three kinds of optimality, and the definitions of generalized forms of convexity and concavity which we shall give in paragraph II,2.

The following definitions of "differential" optimalities are obtained by substituting to the functions f and g their linear approximations at x^o (remember we supposed both f and g to be continously differentiable):

Definition II,1,1'',2'',3''

$x^o \in X$ is differentially (strictly, D.S.O.) (ordinarily, D.O.O.) (weakly, D.W.O.) optimal if no $\bar{x} \neq 0$ exists such that

i) $\nabla f(x^o)\,\bar{x} \in (C)\ (C^1)\ (\text{int } C)$

ii) $\nabla g_I(x^o)\,\bar{x} \geqslant 0$

where $\nabla f(x^o)$ is the matrix whose rows are the gradients of the components of f, and g_I is the vector-valued function whose components are such that $g_i(x^o) = 0$ (the so-called "active" constraints). We shall refer to condition i) by saying that $x^o + \bar{x}$ is "differentially better than x^o" and to condition ii) by saying that it is "differentially feasible". Notice that if $C = N^+$, remembering the definitions of paragraph I,2, the "properly optimal" decisions in the sense of Kuhn and Tucker are the ones which are both L.O.O. and D.O.O. in the above definitions.

Table I summarizes all above defintions of optimality.

TABLE I

	WEAK	ORDINARY	STRICT
GLOBAL	no x^1 exists such that $f(x^1) \gg' f(x^0)$ $g(x^1) \geq 0$	no x^1 exists such that $f(x^1) >' f(x^0)$ $g(x^1) \geq 0$	no $x^1 \neq x^0$ exists such that $f(x^1) \gg' f(x^0)$ $g(x^1) \geq 0$
LOCAL	there is a neighbourhood I of x^0 such that for no $x^1 \in I \cap X$ $f(x^1) \gg' f(x^0)$ $g(x^1) \geq 0$ hold	there is a neighbourhood I of x^0 such that for no $x^1 \in I \cap X$ $f(x^1) >' f(x^0)$ $g(x^1) \geq 0$ hold	there is a neighbourhood I of x^0 such that for no $x^1 \in I \cap X$, $x^1 \neq x^0$ $f(x^1) \gg' f(x^0)$ $g(x^1) \geq 0$ hold
DIFFERENTIAL	no $\bar{x}$ exists such that $\nabla f(x^0)\,\bar{x} \gg' 0$ $g_I(x^0)\,\bar{x} \geq 0$	no $\bar{x}$ exists such that $\nabla f(x^0)\,\bar{x} >' 0$ $g_I(x^0)\,\bar{x} \geq 0$	no $\bar{x}$ exists such that $\nabla f(x^0)\,\bar{x} \gg' 0$ $g_I(x^0)\,\bar{x} \geq 0$

II.2.- SOME GENERALIZED DEFINITIONS OF CONVEXITY AND CONCAVITY

The following Definitions may be considered as extensions to vector functions f, when their range is ordered by the cone V, of the definitions of pseudo-convexity, quasi-convexity, etc... as defined for scalar functions in Mangasarian 1969, ch. 9, to which we constantly refer for terminology.

Although all definitions we shall give are meant to have only local validity, in a neighbourhood of x^o, we shall always understand the term " local" in the denominations we shall use.

Definition II,2,1

The continuously differentiable function h from R^n to R^p is "convex at x^o with respect to a given cone V" of R^p (or, simply, "V-convex at x^o") iff a $\lambda > 0$ exists such that $h(x^o)\, x \in V$ implies $h(x^o + x) - h(x^o) \in V$, for all $x \in \lambda B$, where B is the unit ball of R^n.

Definition II,2,2

The continuously differentiable function h from R^n to R^p is " concave at x^o with respect to a given cone V" of R^p (or, simply, " V-concave at x^o"), iff a $\lambda > 0$ exists such that $\nabla h(x^o)\, x \notin V$ implies $h(x^o + x) - h(x^o) \notin V$, for all $x \in \lambda B$.

The following obvious properties of the two above definitions will be often useful in proving some interesting results: a necessary and sufficient condition for h to be V-convex at x^o is that a $\lambda > 0$ exists such that $h(x^o + x) - h(x^o) \notin V$ implies $\nabla h(x^o)\, x \notin V$ for all $x \in \lambda B$, whereas for V-concavity at x^o, $h(x^o + x) - h(x^o) \in V$ implies $\nabla h(x^o)\, x \in V$ for the some x's.

In the sequel we shall often omit the specification " at x^o", which will be clear from the contest.

In the next sections we shall deal with convexity and concavity of the objective function f with respect to the closed and convex cone C, to C' and to int C. For easy reference we summarize in Table 2 the relations which define the corresponding generalized convexities and concavities if valid for all x lying in a proper neighbourhood of x^o.

TABLE II

with respect to:	CONVEXITY	CONCAVITY
C	$\nabla f(x^o)x \in C \Rightarrow f(x^o+x)-f(x^o) \in C$ or $f(x^o+x)-f(x^o) \notin C \Rightarrow \nabla f(x^o)x \notin C$	$\nabla f(x^o)x \notin C \Rightarrow f(x^o+x)-f(x^o) \notin C$ or $f(x^o+x)-f(x^o) \in C \Rightarrow \nabla f(x^o)x \in C$
C'	$\nabla f(x^o)x \in C' \Rightarrow f(x^o+x)-f(x^o) \in C'$ or $f(x^o+x)-f(x^o) \notin C' \Rightarrow \nabla f(x^o)x \notin C'$	$\nabla f(x^o)x \notin C' \Rightarrow f(x^o+x)-f(x^o) \notin C'$ or $f(x^o+x)-f(x^o) \in C' \Rightarrow \nabla f(x^o)x \in C'$
int C	$\nabla f(x^o)x \in \text{int}\, C \Rightarrow f(x^o+x)-f(x^o) \in \text{int}\, C$ or $f(x^o+x)-f(x^o) \notin \text{int}\, C \Rightarrow \nabla f(x^o)x \notin \text{int}\, C$	$\nabla f(x^o)x \notin \text{int}\, C \Rightarrow f(x^o+x)-f(x^o) \notin \text{int}\, C$ or $f(x^o+x)-f(x^o) \in \text{int}\, C \Rightarrow \nabla f(x^o)x \in \text{int}\, C$

No attempt is made here to build up a general theory of the functions enjoying these properties. We expose now only some relations with usual concepts of convexity and concavity.

If $C = N^+$ then C-convexity reduces to (local) componentwise pseudoconvexity, int C-convexity to (local) componentwise quasiconvexity, whereas C-concavity reduces to (local) componentwise quasiconcavity and int C-concavity reduces to (local) componentwise pseudo concavity (see for example Mangasarian 1969, ch. 9). Obviously, if $C = N = -N^+$, the same reductions as above are produced, in which pseudo- and quasi-convexity are replaced by pseudo- and quasi-concavity, respectively, and viceversa.

II.3.- THE MAIN RESULTS WITHOUT SCALARIZATION

In order to state our sufficient conditions on the functions f and g, which guarantee the equivalence of different kinds of local and differential optimality we shall use the two following Lemmas:

Lemma II.3.1

If there is no $x \neq 0$ such that $\nabla h(x^o)x\varepsilon V$, where h is a continuosly differentiable function from R^n to R^p and V is a closed and convex cone of R^p, then there cannot exist an infinite sequence $\{x^j\}$ such that $x^j \to 0$ and $v^j = h(x^o+x^j)-h(x^o)\varepsilon V$ for all j.

Proof (by contradiction).

Suppose that the infinite sequence $\{x^j\}$ is such that, $x^j \to 0$, and $v^j \varepsilon V$ for all j. By the Bolzano-Weierstrass theorem there exists an infinite subsequence $\{x^k\}$ such that $x^k \to 0$, $\frac{x^k}{\|x^k\|} \to \bar{x}$, $v^k \varepsilon V$ for all k. Consider now the sequence $\left\{\frac{v^k}{\|x^k\|}\right\}$: by the continuous differentiability of h, we may express it as $\left\{\nabla h(x^o)\frac{x^k}{\|x^k\|} + \frac{O(x^k)}{\|x^k\|}\right\}$, with the second term tending to zero as x^k tends to zero. We therefore get $\frac{v^k}{\|x^k\|} \to \nabla h(x^o)\bar{x}$, thus contradicting the hypothesis, since from the fact that $\frac{v^k}{\|x^k\|} \varepsilon V$ for all k we conclude that $\nabla h(x^o)\bar{x}\, \varepsilon V$.

Lemma II.3.2

If there does not exist any infinite sequence $\{x^j\}$ such that $x^j \to 0$ and $v^j = h(x^o+x^j)-h(x^o)\varepsilon V$ for all j, then no $x \neq 0$ exists such that $\nabla h(x^o)x\varepsilon V$, where h is a continuously differentiable and V-convex function at x^o from R^n to R^p, and V is a closed and convex cone of R^p.

Proof (by contradiction).

It suffices to choose $x^j = \frac{1}{j} x$, with x such that $\nabla h(x^o)x\epsilon V$. Then V-convexity implies $v^j \epsilon V$ for all j large enough, a contradiction.

As a consequence of the above lemmas, we have the following theorems:

Theorem II,3,1 (D.S.O. $\Rightarrow$ L.S.O.)

Differential strict optimality implies local optimality, of any kind.

Proof.

It suffices to show that strict local optimality is implied, and this may be done by using Lemma II,3,1 with $p = m + k$, $h = \begin{bmatrix} f \\ g \end{bmatrix}$ and $V = \{v = (y,z) \mid y\epsilon C \subset R^m,\ z\epsilon P^k \subset R^k\}$, with $P^k = \{z \mid z\epsilon R^k,\ z \geq 0\}$.

Theorem II,3,2 (L.S.O. $\Rightarrow$ D.S.O.)

Strict local optimality at x^o implies differential optimality at x^o, of any kind, provided that f is C-convex at x^o and g is pseudo-convex at x^o.

Proof.

The proof is immediate, by using Lemma III,3,2 with the same techniques of the proof of Theorem II,3,1.

Now we consider the case of weak optimality. We state the two following theorems:

Theorem II,3,3 (D.W.O. $\Rightarrow$ L.W.O.)

If x^o is a differentially weakly optimal decision, f is int C-concave at x^o and g is quasi concave at x^o, then x^o is also a locally weakly optimal decision.

Proof.

The last part of the theorem follows by adapting Theorem II,3,1.

to g. The rest od the proof may be carried out by contradiction, by recalling that $f(x^o+x)-f(x^o) \in \text{int } C$ implies $\nabla f(x^o)\, x \in \text{int } C$, since f is int C-concave.

Theorem II,3,4 (L.W.O.$\Rightarrow$ D.W.O.)

If x^o is a locally weakly optimal decision, and g is pseudo-convex at x^o, then x^o is also a differentially weakly optimal decision.

Proof.

The condition on g follows from Lemma III,3,2. That no condition is required on f may be proven by contradiction: choose any sequence $\{x^k\}$ converging to zero with $\frac{x^k}{\|x^k\|} \to \bar{x}$, where $\bar{x}$ is such that $\nabla f(x^o)\bar{x} \in \text{int } C$. Then the sequence $\frac{y^k}{\|x^k\|} = \frac{f(x^o+x^k)-f(x^o)}{x^k}$ tends to $\nabla f(x^o)\bar{x} \in \text{int } C$, thus contradicting the fact that $\frac{y^k}{\|x^k\|}$, and therefore its limit, cannot lie in int C for all k.

For the case of ordinary optimality we use similar arguments to obtain the following theorems:

Theorem II,3,5 (D.O.O.$\Rightarrow$ L.O.O.)

If x^o is a differentially ordinarily optimal decision, f is C'-concave at x^o, and g is quasi-concave at x^o, then x^o is also a locally ordinarily optimal decision.

Proof.

As in Theorem II,3,3 the last part of the statement follows from Theorem II,3,1, and the rest of the proof follows by contradiction, since $f(x^o+x)-f(x^o) \in C'$ implies $\nabla f(x^o)x \in C'$, by the C'-concavity of f at x^o.

Theorem II,3,6 (L.O.O.$\Rightarrow$ D.O.O.)

If x^o is a locally ordinarily optimal decision, f is C'-convex at x^o, and g is pseudo-convex at x^o, then x^o is also a differentially ordinary optimal decision.

Proof.

As in Theorems III,3,2 and III,3,4, the condition on g follows from Lemma III,3,2. The condition on f is proven by contradiction: choose the sequence $\{x^k\}$ with $x^k = \frac{1}{k}\bar{x}$, where $\bar{x}$ is such that $\nabla f(x^o)\bar{x} \in C'$. Then $f(x^o+x^k)-f(x^o) \in C'$ for all k, by the D'-convexity of f at x^o, thus contradicting the ordinary local optimality of x^o.

Notice that pseudo-convexity at x^o is the only " constraint qualification" required by the " even-numbered" theorems (that is, the theorems stating implications from local to differential optimality). Such a general property trivially follows from Lemma III,3,2, as it has been observed. Furthermore, no condition on constraints is required by the " odd-numbered" theorems (that is, the ones stating implications from differential to local optimality). This fact is one of the most relevant consequences of Lemma III,3,1.

II.4.- APPROPRIATE SCALARIZATION, AND SUFFICIENT CONDITIONS FOR DIFFERENTIALLY OPTIMAL DECISIONS TO BE ALSO LOCALLY OPTIMAL

As it is well known, if f(X), or f(X) + D, see Yu 1972, and Yu this volume, is strictly convex, the decisions x_α which are non-dominated with respect to the cone D are characterized by the fact that the scalar form $\sum_1^m \alpha_i f_i(x)$ assumes there its global (local) maximum, when $\alpha \in D^o$, D^o the dual of D. See also Marzollo, Ukovich 1974 and 1975 for the case f(x) + D only convex.

The reason why the above procedure fails to give all globally (locally) non dominated decision without the above mentioned convexity assumptions may be seen as a consequence of the fact that through the scalarization a complete quasi-ordering (+) is introduced on all

(+) We follow here the terminology of Sen, 1970.

$x \in X$ (or a neighbourhood of x_α) so that, for each $\alpha \in D^o$, x' is "better or equal" x" iff $\Sigma_{1i}^m \alpha_i f_i (x') \geq \Sigma_{1i}^m \alpha_i f_i (x'')$. Therefore comparisons are introduced among decisions which may be not comparable in the original problem, for example between two decisions which are both Pareto-optimal, in the case D = N.

Dealing with local Pareto-optimality and referring to the case of non convex f (X) + N, Wan (see Theorem 2, Lemma 1 and Theorem 3 in Wan 1975) recently extended some previous results of Smale (see Smale 1973 and 1974) by giving a second order condition (see Wan 1975, Theorem 2, Lemma 1 and Theorem 3) which we restate here, using our terminology: a differentially weakly Pareto-optimal decision x^o is also locally strictly Pareto-optimal if, for all $x \neq 0$ such that $\nabla f(x^o)\, x > 0$, the quadratic form $x, \nabla^2 \bar{y} f (x^o)\, x$ is negative, where $\bar{y} > 0$ is such that $\bar{y}\, \nabla f(x^o) = 0$ and $\nabla^2 \bar{y} f$ is the hessian matrix of the scalar function $\bar{y} f$ at x^o.

Using a different approach from Smale and Wan, we shall here generalize the previous sufficient conditions for local optimaly in two relevant directions, since we shall deal with non-dominacy with respect to a broad classes of (possibly non-polyhedrical) cones instead of the only Pareto-optimality, and since the condition we shall put on the scalarized function is implied by the previous second order condition.

As in Wan 1975, we shall deal here only with the unconstrained case $X = R^n$, but the extension to the constrained case $X = \{x \mid g(x) \geq 0\}$ is easy and follows the same pattern as in Marzollo, Serafini, Ukovich 1976, where Pareto-optimality was treated.

In the following theorems, which will relate differential to local optimalities, we shall need appropriate versions of some so called "separation" or "alternative" theorems, which we state in the form of Lemma II,4,1 and Lemma II,4,2 . (For an exhaustive review of the "theorems of the alternative" dealing with mutual incompatibility of the feasibility of two systems of linear inequalities, see for example Gale 1960 and Mangasarian 1969. For more sophisticated versions, see Berge and Gouila Houri 1968, Rockafellar 1970, Hoang 1974, Hurwicz 1958).

Lemma II,4,1

If the range of a given $(m \times n)$ matrix A does not meet the interior of a given (full-dimensional) convex cone C of R^m, then there exists a non zero vector $\bar{y}$ in $C^{oo}=\{y \mid y \in R^m,\ y\,y' \geq 0 \text{ for all } y' \in C\}$ the polar cone of C, such that $\bar{y}\,A = 0$.

Proof.

The conditions of Theorem 11.2 of Rockafellar 1970 are met, hence a hyperplane containing the range of A exists, which does not meet int C. Then $\bar{y}$ is any normal to such a hyperplane pointing toward the halfspace containing int C, as it may be easily checked.

Lemma II,4,2

If the intersection of the range of a given $(m \times n)$ matrix A with a given (full-dimensional) convex cone C of R^m contains the only origin, then there exists a vector $\bar{y}$ in int $C^{oo} = \{y \mid y \in R^m,\ y\,y' > 0, \text{ for all } y' \in C\}$ such that $\bar{y}\,A = 0$.

Proof.

By the "main separation theorem" (Theorem 11.3 of Rockafellar 1970) there exists a hyperplane properly separating C from the range of A. Then again $\bar{y}$ is any normal to such a hyperplane pointing toward the (closed) halfspace containing C.

We state now the main results of this paragraph:

Theorem II,4,1 (D.W.O. $\Longrightarrow$ L.S.O.)

If x^o is a differentially weakly optimal decision, a nonzero vector $\bar{y}$ exists in C^{oo} such that $\bar{y}\ \nabla f(x^o) = 0$. If there exists an open cone B_2 in R^n and a neighbourhood I of the origin such that

$$B_1' \subset B_2' \subset B_3', \qquad \text{II,4(1)}$$

where,

$B_1' = I \cap B_1$; $B_1 = \left\{ x \mid x \neq 0, \nabla f(x^o)\, x \in C \right\}$, $B_2' = I \cap B_2$,

$B_3' = I \cap B_3$; $B_3 = \left\{ x \mid \bar{y}\ (f(x^o + x) - f(x^o)) < 0 \right\}$,

then x^o is also a locally strictly optimal decision (hence also locally ordinarily and weakly optimal).

Proof.

The first part of the statement simply follows from Lemma II,4,1 by taking $\nabla f(x^o) = A$, since by weak differential optimality $\nabla f(x^o)x \notin \text{int } C$, for all x.

The second part is proven by contradiction: suppose that x^o is not a locally strictly optimal decision; then there exists a nonzero infinite sequence $\left\{x^j\right\}$ tending to zero such that, for all j,

$$f\ (x^o + x^j) - f(x^o) \in C.$$

By the Bolzano-Weierstrass theorem, there exists an infinite subsequence $\left\{x^k\right\}$ such that $\frac{x^k}{\|x^k\|} \longrightarrow \bar{x}$.

Clearly, $x^k \notin B_3$ for all k, by definition of B_3 and since $\bar{y} \in C^{oo}$, $f(x^o + x^k) - f(x^o) \in C$. Hence $x^k \notin B_2'$ for sufficiently large k, and therefore $\frac{x^k}{\|x^k\|} \notin B_2$, for sufficiently large k. But $\bar{x} \in B_1$, since $\frac{f(x^o+x^k)-f(x^o)}{\|x^k\|}$ is in the closed cone C for all k, and it tends to $f(x^o)\ \bar{x}$. We thus have a contradiction, since our hypothesis implies $B_1 \subset B_2$.

It may be noticed that condition II.4(1) in the above theorem represents a very weak kind of concavity at x^o of the function $\bar{y}$ f, <u>since it is required only for some special directions</u>. If one recalls the following definition (see Ponstein 1967): a real valued function h is strictly pseudoconcave at x^o iff $h(x^o+x)-h(x^o) \geq 0$ implies $\nabla h(x^o)x > 0$ (or, equivalently, $\nabla h(x^o)x \leq 0$ implies $h(x^o+x)-h(x^o) < 0$), the above condition may be seen as "local strict C-pseudoconcavity" required only in the directions x such that $\nabla f(x^o)\, x \in$ boundary of C (by differentially weak optimality there is no x such that $\nabla f(x^o)\, x \in$ int C). Remark also that the esistance of the open cone B_2 is essential for the validity of Theorem II,4,1; take for example

$n = 2, \quad m = 3 \quad C = N^+$

$f_1(x) = x_1^2 - x_2$

$f_2(x) = x_2 - x_1^2 \qquad\qquad x^o = \begin{matrix} 0 \\ 0 \end{matrix}$

$f_3(x) = x_2^2 - x_1^4$

$$\nabla f(x^o) = \begin{vmatrix} 0 & -1 \\ 0 & 1 \\ 0 & 0 \end{vmatrix}$$, so we may choose $\bar{y} = (1, 1, 1)$, hence

$\bar{y}\, f(x) = x_2^2 - x_1^4$. Clearly, $B_1 = \left\{ (x_1, 0) \mid x_1 \in R^1 \right\}$ is contained in $B_3 = \left\{ x \mid -x_1^2 < x_2 < x_1^2 ,\; x_1 \in R^1 \right\}$, but no B_2 may be found, and actually the origin is not a locally Pareto-optimal decision.

From the computational point of view, the following Corollary is useful:

<u>Corollary II,4,1</u>

If x^o is a differentially weakly optimal decision, and $x \in B_1$ implies $x\, \nabla^2 \bar{y}\, f(x^o)\, x < 0$, $x \neq 0$, where $\nabla^2 \bar{y}\, f(x^o)$ is the hessian matrix of $\bar{y}\, f$ at x^o, with $\bar{y} \in C^{oo}$ such that $\bar{y}\, \nabla f(x^o) = 0$, then x^o is also a locally strictly optimal decision.

Proof.

Simply let B_2 be $\left\{ x \mid x\, \nabla^2 \bar{y}\, f(x^o)\, x < 0 \right\}$, and the conditions of Theorem II,4,1 are met.

Notice that Corollary II,4,1 extends Theorem 3 in Wan, 1975.

With the same arguments as in Theorem II,4,1 and using Lemma II,4,2 instead of Lemma II,4,1, it is also possible to give sufficient conditions for differentially ordinarily weakly optimal decisions to be also locally ordinarily (weakly) optimal. We state these conditions in the following theorems.

Theorem II,4,2

If x^o is a differentially ordinarily optimal decision, a non-zero vector $\bar{y}$ exists in int C^{oo} such that $\bar{y}, \nabla f(x^o) = 0$. Repeating the same hypotheses an in Theorem II,4,1 with the only modification of B_3 sostituted with $\left\{ x \mid \bar{y}, \left[f(x^o + x) - f(x^o) \right] \leq 0 \right\}$, x^o is a also locally ordinarily optimal decision.

Theorem II,4,3

Theorem II,4,2 holds if we substitute the word "ordinary" with the word "weakly". Notice that the "local strict C pseudoconcavity" required in Theorem II,4,1 is relaxed in the last theorems, since only "local C concavity" is now required.

II.5.- CONCLUSIONS

In Part II,1, nine kinds of optimalities were defined with respect to a cone C. These optimalities reduce to what we called weak, ordinary or strict Pareto optimality when the cone C is int N^+, $N^+/0$, N^+ respectively. With the definition in II,2 of appropriate types of convexity with respect to a cone, a complete set of sufficient conditions was given on the behaviour of the objective function f and the constraint function g under which each one of the defined optimalities implies each other one. For the particular case $C = N^+/0$, (and for polyedrical cones, a situation which may easily be reduced to the Pareto case) some of these conditions were already available in the literature, some are new; in general an effort was made to weaken as much as possible the hypotheses on f and g, dealing in a unitary way with polyedrical or non-polyedrical cones.

Referring to polyedrical cones, hence to Pareto optimalities, in Theorem II,4,1 we required only strict pseudoconcavity of the linear combination $\bar{y}$,f of the components f_i (which is obviously less stringent than pseudoconcavity of each f_i), limited to some special directions, in order to infer locally strict from differentially weak optimality, and strict pseudoconcavity was relaxed into concavity of $\bar{y}$,f in order to infer local ordinary (weak) optimality from differential ordinary (weak) optimality.

The results of II,4 are important also from a practical point of view, since the very definition of differential optimalities allows an easy way of checking them, and the conditions for passing from differential to local optimalities suggest an obvious way of testing them.

The check for differential optimalities actually consists in verifying the unfeasibility of the following linear systems of inequalities (unconstrained case)

$$\nabla f(x^o)\, x \geqslant 0 \qquad \text{II,5(1)}$$
$$x \neq 0$$

for differentially strict optimality;

$$\nabla f(x^o)\, x > 0 \qquad \text{II,5(2)}$$

for differentially ordinary optimality;

$$\nabla f(x^o)\, x \gg 0 \qquad \text{II,5(3)}$$

for differentially weak optimality.

The above verification may efficiently be carried out by adapting the simplex method to the case of absence of an objective function (see Marzollo, Serafini, Ukovich 1976). If II,5(1) is unfeasible, using Theorem II,3,1 with, $C = N^+$, we may also state locally strict (hence also ordinary and weak) optimality of x^o.

If II,5(1) is feasible but II,5(2) is not, we have to solve the linear system $\bar{y}, \nabla f(x^o) = 0$ obtaining an $\bar{y} \gg 0$, and then to verify whether the conditions of Theorem II,4,1 are satisfied, where B_1 reduces to the kernel of $\nabla f(x^o)$: $\{ x \mid \nabla f(x^o)\, x = 0 \}$. If they are, x^o is locally strictly optimal. If they are not, but the condition of Theorem II,4,2 are satisfied (with B_1 given again by the kernel of $\nabla f(x^o)$), x^o is locally ordinarily optimal.

If II,5(1) and II,5(2) are feasible, but II,5(3) is not, the solution of the linear system $\bar{y}, \nabla f(x^o) = 0$ gives us an $\bar{y} > 0$. In correspondence to such $\bar{y}$, if the conditions of Theorem II,4,1 are satisfied (with B_1 given by the cone $\{ x \mid \nabla f(x^o)\, x \geq 0 \}$), x^o is locally strictly optimal. If they are not, but those of Theorem II,4,3 are, x^o is locally weakly optimal.

Needless to say, series expansions of $\bar{y}$, f around x^o are a practical way of testing the above conditions, as it was done in Corollary II,4,1.

For the constrained case, the condition $\nabla g_I(x^o)\, x \geq 0$ is added to the systems II,5,(1)(2)(3), which may be verified again with the algorithm suggested in Marzollo, Serafini, Ukovich 1976; and also the verification of the conditions which result from the analogues of Theorems II,4,1,2,3 (with the scalar function $\bar{y},f + \bar{z},g$ substituting $\bar{y},f$, etc.) follows the same patterns.

BIBLIOGRAPHY

J.P. Aubin: "A Pareto Minimum Principle", in Kuhn, Szegö (eds.): "Differential Games and Related Topics", North-Holland, 1971.

C. Berge, A. Gouila-Houri: "Programming, Games and Transportation Networks", Wiley, 1965.

A. Blaquière: "Vector Valued Optimization in Multiplayer Quantitative Games", this volume, pp. 33 - 54, 1976.

A. Blaquière, L. Juricek, K. Wiese: "Geometry of Pareto Equilibria and Maximum Principle in N-Person Differential Games" J. of Math. An. and Appl. vol. 38, 1972.

G. Birkhoff: "Lattice Theory" American Mathematical Society Colloquium Pubblications vol. XXV, 1948.

S.S.L. Chang: "General Theory of Optimal Processes" J. SIAM on Control vol. 4, pp. 46 - 55, 1966.

K.C. Chu: "On the Noninferior Set for the Systems with Vector-Valued Objective Function" IEEE Tr. on Aut. Contr. AC-15, pp. 103-124, 1970.

N.O.Da Cunha, E. Polak: "Constrained Minimization Under Vector-Valued Criteria in Finite Dimensional Spaces" J. of Math. An. and Appl. vol. 19, pp. 103 - 124, 1967.

G. Debreu: "Theory of Value" Wiley, 1959.

D. Gale: "The Theory of Linear Economic Models", McGraw-Hill, 1960.

A.M. Geoffrion: "Strictly Concave Parametric Programming, Part I: Basic Theory", Management Science, vol. 13, pp. 244 - 253,1966.

A.M. Geoffrion: "Strictly Concave Parametric Programming, Part II: Additional Theory and Computational Considerations", Management Science, vol. 13, pp. 359 - 370, 1967a.

A.M. Geoffrion: "Solving Bicriterion Mathematical Programs" Op.Res., vol. 15, pp. 39 - 54, 1967b.

A. Haurie: "On Pareto Optimal Decisions for a Coalition of a Subset of Players" IEEE Tr. on Aut. Contr. AC-18, pp. 144 - 149, 1973.

A. Haurie, M.C.Delfour: "Individual and Collective Rationality in a

Dynamic Pareto Equilibrium", J.O.T.A. vol. 13, pp. 290 - 302, 1974.

T. Hoang: "The Farkas-Minkowski Theorem and Extremum Problems", in J. and M. Łos (eds.): "Mathematical Models in Economics", North-Holland, 1974.

L. Hurwicz: "Programming in Linear Spaces", in K.J. Arrow, L. Hurwicz, H. Uzawa (eds.): "Studies in Linear and Nonlinear Programming", Stanford Univ. Press, 1958.

S. Karlin: "Mathematical Methods and Theory in Games, Programming and Economics" vol. 1. Addison-Wesley, 1959.

R.L. Keeney, H. Raiffa: "Decision Analysis with Multiple Conflicting Objectives, Preferences and Value Tradeoffs" II ASA Working Paper WP - 75 - 73, 1975.

A. Klinger: "Vector-Valued Performance Criteria" IEEE Tr. on Aut. Contr. AC-9, pp. 117 - 118, 1964.

A. Klinger: "Improper Solutions of the Vector Maximum Problem" Op. Res. vol. 15, pp. 570 - 572, 1967.

T.C. Koopmans (ed.): "Activity Analysis of Production and Allocation", Cowles Commission Monograph N° 13, Wiley, 1951.

H.W. Kuhn, A.W. Tucker: "Nonlinear Programming" in J. Neyman (ed.): Proc. of the Second Berkeley Symposium on Mathematical Statistics and Probability, Univers. of Calif. Press, pp. 481 - 491, 1951.

G. Leitmann, S. Rocklin, T.L. Vincent: "A Note on Control-Space Properties of Cooperative Games", J.O.T.A. vol. 9, 1972.

G. Leitmann, W. Schmitendorf: "Some Sufficiency Conditions for Pareto-Optimal Control" J. of Dynamic Systems, Measurement and Control, vol. 95, pp. 356 - 361, 1973.

G. Leitmann: "Cooperative and Non-Cooperative Many Player Differential Games", Springer Verlag, 1974.

G. Leitmann: "Cooperative and Non-Cooperative Differential Games", this volume, pp. 7 - 32, 1976.

R.D. Luce, H. Raiffa: "Games and Decisions" Wiley, 1957.

K.R. Mac Crimmon: "An Overview of Multiple Objective Decision Making" in J.L. Cochrane, M. Zeleny (eds.): "Multiple Criteria Decision Making", Univers. of Southern Calif. Press, pp. 18 - 44,

1973.

O. Mangasarian: "Nonlinear Programming" McGraw-Hill, 1969.

A. Marzollo, W. Ukovich: "A Support Function Approach to the Characterization of the Optimal Gain Vectors in Cooperative Games" Proc. of the 1974 IEEE Conference on Decision and Control, pp. 362 - 367, 1974.

A. Marzollo, W. Ukovich: "Nondominated Solutions in Cooperative Games: A Dual Space Approach", Proc. of the VI IFAC Congress, Boston, 1975.

A. Marzollo, P. Serafini, W. Ukovich: "Decisioni Paretiane e loro determinazione", in "Teoria dei Sistemi ed Economia" GES - Il Mulino, 1976.

W. Pareto: "Cours d'Economie Politique" Ronge, 1896.

E. Polak, A.N. Payne: "On Multicriteria Optimization" ERL - M 566 Univ. of Californ., Berkeley, May 1975.

J. Ponstein: "Seven Kinds of Convexity", SIAM Review, vol. 9, N° 1, 1967.

E.V. Rekasius, W.E. Schmitendorf: "On the Noninferiority of Nash Equilibrium Solutions" IEEE Tr. on Aut. Contr. AC - 16, pp. 170 - 173, 1971.

R.T. Rockafellar: "Convex Analysis", Princeton Press, 1970.

B. Roy: "Problems and Methods with Multiple Objective Functions", Math. Progr., vol. 1, pp. 239 - 266, 1971.

W.E. Schmitendorf: "Cooperative Games and Vector-Valued Criteria Problems" IEEE Tr. on Aut. Contr. AC - 18, pp. 139 - 144, 1973.

A.K. Sen: "Collective Choice And Social Welfare" Oliver and Boyd, 1970.

S. Smale: "Global Analysis and Economics I: Pareto Optimum and a Generalization of Morse Theory", in M. Peixoto (ed.): Dynamical Systems, pp. 531 - 544, Academic Press, 1973.

S. Smale: "Global Analysis and Economics V: Pareto Theory with Constraints" J. of Math. Economics, vol. 1, pp. 213 - 221, 1974.

W. Stadler: "Preference Optimality and Applications of Pareto Optimality", this volume.

A.W. Starr, Y.C. Ho: "Nonzero-Sum Differential Games" J.O.T.A., vol. 3, pp. 184 - 206, 1969a.

A.W. Starr, Y.C. Ho: "Further Properties of Nonzero-Sum Differential Games" J.O.T.A., vol. 3, pp. 207 - 219, 1969b.

T.L. Vincent, G. Leitmann: "Control-Space Properties of Cooperative Games" J.O.T.A., vol. 6, pp. 91 - 104, 1970.

Y.H. Wan: "On Local Pareto Optimum" J. of Math. Economics, vol. 2, 1975.

P.L. Yu: "Cone Convexity, Cone Extreme Points and Nondominated Solutions in Decision Problems with Multiobjectives", University of Rochester, Center for System Science, Report 72 - 02, 1972, reprinted in J.O.T.A., 1974.

P.L. Yu: "Domination Structures and Nondominated Solutions", this volume.

L.A. Zadeh: "Optimality and Non-Scalar-Valued Performance Criteria" IEEE Tr. on Aut. Contr. AC - 8, pp. 59 - 60, 1963.

ESTIMATING THE COMMON COST OF A GOOD WHEN THE LOCAL COSTS ARE KNOWN IN THE COUNTRIES OF A COMMUNITY

Mario Volpato
Department of Mathematics
University of Venice

1. INTRODUCTION.

In an economic community among many countries, the estimate of the common cost of any good, when the local costs in the countries are known, is of substantial interest also for pure theory. In fact, the classical point of view of the compared costs solves the problem under the hypothesis that the local global costs are linear, which involves that the local costs are independent of the quantity of the good. It is also limited (as far as we know) to the case of only two countries.

In this paper we consider the problem with any number of partners and assuming that the local global costs are not necessarily linear.

2. A CRITERION FOR THE ESTIMATE OF THE COMMON GLOBAL COST.

Starting from the classical theory of the compared costs, our criterion to determine the common global cost of any amount X_h (of any good h) makes use of an estimate of the penalty induced by the production of X_h on the production of any other good k, to be assumed as standard. In other words, the global common cost of the amount X_h of h is given by the quantity Z_k (of the standard good k) that the community cannot produce since X_h is produced. The criterion is not yet defined. Indeed, the amount Z_k which is not produced depends on the policy of the community, dividing among the countries the production of the total amount X_h. The obvious partition is the one minimizing the common penalty Z_k. This is achieved by a policy (to be called optimal, or community-optimal) assigning to each country the production of a part of X_h (with the constraint of the total amount X_h) in such a way that the (possible) residual resources, if fully used for the production of the standard good k, give the maximum yield. The criterion for the estimate of the common global cost is now defined, it consists in solving the problem of optimizing the production.

3. PRODUCTION ABILITY OF EACH COUNTRY.

Let $P_1, P_2, \ldots, P_n$ be the countries of the community and assume that any country P_i can direct to the production of the good h certain resources that can produce at most the amount R_{ih} of h. It follows that the maximum amount of h that the community could produce is given by the sum

$$\sum_{i=1}^{n} R_{ih} \overset{\text{def}}{\equiv} R_h \ . \tag{1}$$

The community production X_h of h can then vary in the range

$$H \overset{\text{def}}{\equiv} 0 \leqslant X_h \leqslant R_h \ . \tag{2}$$

We will better think of this range as the addition of the n adjacent subranges

$$I_{s+1,h} \overset{\text{def}}{\equiv} \sum_{i=1}^{s} R_{ih} \leqslant X_h \leqslant \sum_{i=1}^{s+1} R_{ih} \ , \tag{3}$$

$$(s = 0, 1, \ldots, n-1; \ \sum_{i=1}^{0} R_{ih} \overset{\text{def}}{\equiv} 0) \ .$$

Each one can again be split into three sets according to

$$I_{s+1,h} = (\bar{H}_s \overset{\text{def}}{\equiv} \sum_{i=1}^{s} R_{ih}) + (H_{s+1} \overset{\text{def}}{\equiv} \sum_{i=1}^{s} R_{ih} < X_h < \sum_{i=1}^{s+1} R_{ih}) +$$

$$+ (\bar{H}_{s+1} \overset{\text{def}}{\equiv} \sum_{i=1}^{s+1} R_{ih}) \ . \tag{4}$$

Here H_{s+1} is the internal of $I_{s+1,h}$; $\bar{H}_s$ and $\bar{H}_{s+1}$ are degenerate subranges (whence the bar), respectively the left and right extrema of $I_{s+1,h}$

The range of $I_{s+1,h}$ (i.e. $R_{s+1,h}$) is clearly the amount of h that the country P_{s+1} can produce. Using the known symbols, the equality holds

$$H = \bigcup_{s=0}^{n-1} \left\{ I_{s+1,h} = \bar{H}_s + H_{s+1} + \bar{H}_{s+1} \right\} \ . \tag{5}$$

Assume then that the same resources which could be used in the various

countries to produce h can be converted to the production of another good k to be conventionally used as standard. More precisely, assume that in the country P_i, the total conversion of the resources useful to produce h gives a yield R_{ik} of k.
The maximum production of k in the community would then be

$$\sum_{i=1}^{n} R_{ik} \overset{\text{def}}{\equiv} R_k . \tag{6}$$

Then the community production Y_k of k can vary in the range

$$K \overset{\text{def}}{\equiv} 0 \leqslant Y_k \leqslant R_k . \tag{7}$$

Again, K is the union of n adjacent subranges whose amplitude is the maximum affordable production of k in each country. It will be useful, below, to write now in the first place the range whose amplitude gives the maximum production in the country P_n, in the second P_{n-1} and so on. In symbols

$$K = \bigcup_{t=0}^{n-1} \left\{ J_{n-t,k} = \bar{K}_{n-t+1} + K_{n-t} + \bar{K}_{n-t} \right\} , \tag{8}$$

where

$$J_{n-t,k} \overset{\text{def}}{\equiv} \sum_{i=n-t+1}^{n} R_{ik} \leqslant Y_k \leqslant \sum_{i=n-t}^{n} R_{ik} = (\bar{K}_{n-t+1} \overset{\text{def}}{\equiv} \sum_{i=n-t+1}^{n} R_{ik}) +$$

$$+ (K_{n-t} \overset{\text{def}}{\equiv} \sum_{i=n-t+1}^{n} R_{ik} < Y_k < \sum_{i=n-t}^{n} R_{ik}) + (\bar{K}_{n-t} \overset{\text{def}}{\equiv} \sum_{i=n-t}^{n} R_{ik}) ;$$

$$(t = 0, 1, \ldots, n-1; \sum_{i=n+1}^{n} R_{ik} \overset{\text{def}}{\equiv} 0) . \tag{9}$$

4. CONVERSION COSTS.

We assumed in the previous section that in the country P_i the resources producing the amount R_{ih} of h, if properly converted for the production of the standard good k, give a yield R_{ik}. Now, in the classical theory of compared costs, the ratio

$$\frac{R_{ik}}{R_{ih}} \overset{\text{def}}{===} q_{ihk} \, , \tag{10}$$

is assumed as unit yield in the h → k conversion. In other words, the conversion h → k (i.e. from the good h to the good k) of the resources that in the country P_i produce a unit amount of h, should yield q_{ihk} units of the standard good k. This way, in the country P_i, a unit of h should cost as much as q_{ihk} units of k.
As a consequence, q_{ihk} is assumed as the price (in standard units) of the good h, and the product

$$q_{ihk} \cdot x_{ih} \tag{11}$$

as the global cost, in P_i, of the amount x_{ih} of h. The classical theory can conflict with real life. Indeed, it souds reasonable that the price of the conversion h → k can depend on the quantity to be converted, so that the global cost to convert x_{ih} is not linear, as in (11).
For this reason we assume that in the country P_i the resources used to produce any amount x_{ih} of h, if converted to produce the standard good k, give a yield

$$y_{ik} = f_{ihk}(x_{ih}) \quad , \quad \text{with} \quad f_{ihk}(0) = 0 \quad , \tag{12}$$

where $f_{ihk}(x_{ih})$ is a function (not necessarily linear) giving the global cost, in the country P_i, of the amount x_{ih} of h, in standard units. We admit that at the same time, in the same country P_i, the inverse conversion $k \to h$ is possible, in such a way that by converting the amount y_{ik} of k into the good h one has

$$x_{ih} = f_{ikh}(y_{ik}) \ . \tag{13}$$

We were able this way to define the functions $f_{ihk}(x_{ih})$ and $f_{ikh}(y_{ik})$, inverse of each other.
Assuming that they are differentiable, one has

$$f'_{ihk}(x_{ih}) \cdot f'_{ikh}(y_{ik}) = 1 \ , \tag{14}$$

which means that the marginal costs of the inverse conversions $h \to k$ and $k \to h$ are the reciprocal of each other.
One can remark that $f'_{ihk}(x_{ih})$ can be regarded, to a first approximation, as the cost (in standard units) to be paid if the conversion $h \to k$ goes on for a further unit of h, after x_{ih} units have already been converted, and similar statement could be given for $f'_{ikh}(y_{ik})$. It then represents the price of h (in standard units) when the amount x_{ih} has been already produced. Unlike in the classical theory, in our hypothesis, this price can depend on the amount x_{ih} of the good considered. In the classical theory (where the global costs are linear, so that the prices are independent of the quantity) (14) becomes

$$q_{ihk} \cdot q_{ikh} = 1 \ . \tag{15}$$

It is useful to remark that an inequality like

$$f'_{ihk}(x_{ih}) \leqslant f'_{shk}(x_{sh}) \ , \qquad (16)$$

shows that the production of a unit of h (when the production in the country P_i is x_{ih} and in P_s is x_{sh}) costs less (in standard units) in P_i than in P_s. At the assumed level of production, the country P_i is better qualified than P_s to produce h. After (14), (16) becomes

$$\frac{1}{f'_{ikh}(y_{ik})} \leqslant \frac{1}{f'_{skh}(y_{sk})} \ , \qquad (17)$$

i.e.

$$f'_{skh}(y_{sk}) \leqslant f'_{ikh}(y_{ik}) \ . \qquad (18)$$

It is enough to say that (16), assuming the production levels given above, evidentiates, at the same time, the better qualification of P_s with respect to P_i in producing k.
Finally remark that, using the last symbols, the equalities hold

$$f_{ihk}(R_{ih}) = R_{ik}, \ f_{ikh}(R_{ik}) = R_{ih}, \ (i = 1,\ldots, n) \ , \qquad (19)$$

which were implicity assumed in section 3.

5. OPTIMAL PARTITION (AMONG THE COUNTRIES OF THE COMMUNITY) OF THE PRODUCTION OF A FIXED QUANTITY X_h OF h.

The criterion adopted to estimate the global community cost of any quantity $X_h \in H$ of the good h gives a problem of optimal production, as announced in section 2, to be stated now.

- The use of the resources which produce the total amount X_h of h has to be partitioned among the countries of the community, so that the (possible) residual resources of each country, if fully used to produce the standard good k, give the maximum yield.

For the analytical formulation let us remark that the (unknown) amount x_{ih} of h that the country P_i should produce satisfies the relations

$$\sum_{i=1}^{n} x_{ih} = X_h \;, \; 0 \leqslant x_{ih} \leqslant R_{ih} \;, \; (i = 1,\dots, n)$$

whose interpretation is immediate. It follows that, in P_i, the residual resources (which could be used to produce h) can further produce $R_{ih} - x_{ih} \overset{\text{def}}{\equiv} y_{ih}$ units of k. With full conversion to the production of k, the yield $f_{ihk}(R_{ik} - x_{ih})$ is obtained. The analytical problem to be solved is then

$$\begin{cases} \max \sum_{i=1}^{n} f_{ihk}(R_{ih} - x_{ih}) \overset{\text{def}}{\equiv} Y_k(X_h) \\ \sum_{i=1}^{n} x_{ih} = X_h \;, \qquad (0 \leqslant X_h \leqslant R_h), \\ 0 \leqslant x_{ih} \leqslant R_{ih}, \qquad (i = 1,\dots,n), \end{cases} \tag{20}$$

where the maximum and the optimal policy to achieve it clearly depend on X_h.

6. THE LINEAR CASE.

In the linear case the objective function of the problem (20) is

$$\sum_{i=1}^{n} q_{ihk} \cdot (R_{ih} - x_{ih}) ,$$

and then, according to the evident relations

$$\max \sum_{i=1}^{n} q_{ihk} \cdot (R_{ih} - x_{ih}) = \sum_{i=1}^{n} q_{ihk} R_{ih} - \min \sum_{i=1}^{n} q_{ihk} x_{ih} =$$
$$= \sum_{i=1}^{n} R_{ik} - \min \sum_{i=1}^{n} q_{ihk} x_{ih} , \qquad (21)$$

the following problem is obtained

$$\begin{cases} \min \sum_{i=1}^{n} q_{ihk} x_{ih} \\ \sum_{i=1}^{n} x_{ih} = X_h , \quad (0 \leqslant X_h \leqslant R_h) , \\ 0 \leqslant x_{ih} \leqslant R_{ih}, \quad (i = 1,\dots,n) . \end{cases} \qquad (22)$$

Without loss of generality it can be assumed that the symbols are chosen so that the inequalities

$$q_{1hk} \leqslant q_{2hk} \leqslant \dots \leqslant q_{ihk} \leqslant \dots \leqslant q_{nhk} , \qquad (23)$$

are satisfied. This means that, according to (16), the countries with a lower numeration index are qualified to the production of the h good and the ones with higher numeration index are qualified to the production of the standard k.

In (22) some known inequalities [1] and the constraint $\sum_{i=1}^{n} x_{ih} = X_h$, give

$$q_{1hk} X_h \leqslant \frac{q_{1hk} x_{ih} + q_{2hk} x_{2h}}{x_{1h} + x_{2h}} X_h \leqslant \dots \leqslant \frac{\sum_{i=1}^{s} q_{ihk} x_{ih}}{\sum_{i=1}^{s} x_{ih}} X_h \leqslant$$

$$\leqslant \frac{\sum_{i=1}^{s+1} q_{ihk} x_{ih}}{\sum_{i=1}^{s+1} x_{ih}} X_h \leqslant \dots \leqslant \frac{\sum_{i=1}^{n} q_{ihk} x_{ih}}{\sum_{i=1}^{n} x_{ih}} X_h =$$

$$= \sum_{i=1}^{n} q_{ihk} x_{ih} . \tag{24}$$

These are sufficient to claim that, if the quantity X_h is not greater, as an example, than the bound $\sum_{i=1}^{s} R_{ih}$ (i.e. the maximum which can be produced with the integral exploitation of the resources of the first s countries) then the optimal policy (in this case unique) does not assign any production of the h to the countries which follows P_s. Precisely, if $X_h \in I_{s+1,h}$, then the optimal policy splits the production in the following way

$$x_{1h} = R_{1h};\ x_{2h} = R_{2h};\ \dots ;\ x_{sh} = R_{sh};\ x_{s+1,h} = X_h - \sum_{i=1}^{s} R_{ih};$$

$$x_{s+2,h} = 0;\ \dots ;\ x_{nh} = 0 . \tag{25}$$

The corrispondent maximum of the optimization problem, for (21) is

$$\sum_{i=s+1}^{n} R_{ik} - q_{s+1,hk} (X_h - \sum_{i=1}^{s} R_{ih}) \overset{def}{=\!=\!=} Y_k ,\ (X_h \in I_{s+1,h}) ; \tag{26}$$

it can be obtained with the integral exploitation, in the production of the standard good k, of the residual resources of the different countries. If these are expressed in units of h, they are distributed in the following way

$$y_{1h} = 0;\ y_{2h} = 0;\ \dots;\ y_{sh} = 0;\ y_{s+1,h} = \sum_{i=1}^{s+1} R_{ih} - X_h;$$

$$y_{s+2,h} = R_{s+2,h};\ \dots;\ y_{nh} = R_{nh}\ . \tag{27}$$

In the conversion $h \rightarrow k$, the quantities

$$y_{1k} = 0;\ y_{2k} = 0;\ \dots;\ y_{sk} = 0;\ y_{s+1,k} = q_{s+1,hk}\left(\sum_{i=1}^{s+1} R_{ih} - X_h\right)$$

$$= R_{s+1,k} - q_{s+1,hk}\,(X_h - \sum_{i=1}^{s} R_{ih});\ y_{s+2,k} = R_{s+2,k};\ \dots;\ y_{nk} = R_{nk}, \tag{28}$$

are obtained, with sum equal to Y_k.
The optimal policy indicates how to assigne the resources of the different countries to the production of the quantity X_h of h and of the correspondent quantity of the standard item k. It exploits, as one could expect, as much as possible the qualification of the different countries, given by the (23), in the production of the goods h and k.

7. THE NON LINEAR CASE.

Generally an optimal policy can't be explicitly written for the problem (20) (as in the linear case). It can be obtained when the conversion costs have the following expressions:

$$f_{ihk}(x_{ih}) = a_{ihk} + b_{ihk}\, x_{ih} + c_{ihk}\, x_{ih}^2 , \quad (b_{ihk} > 0,\ c_{ihk} < 0) ,$$ [2]

$$f_{ihk}(x_{ih}) = a_{ihk}\,(1 - e^{-b_{ihk} x_{ih}}) ,$$ [3]

$$f_{ihk}(x_{ih}) = \frac{a_{ihk}\, x_{ih}}{b_{ihk}\, x_{ih} + c_{ihk}} ,$$ [4]

as it can be expected looking at the results obtained in certain problems, analitically quite similar, by O. Cucconi [2], C. Wilkinson - S.K. Gupta [3], G. Castellani [4] and, to a certain extent, in the contribution to this volume by the latest author, where the analytical problem is extended to the case of any convex function, thus including the three kinds considered above. When the functions representing the conversion costs are (even only semicontinuous but) tabulated at least for some discrete values of their argument, the solution can be found by acting step by step using dynamic programming [5].

We will now demonstrate that if, in the various countries, the costs of the conversion $h \to k$ satisfy in all points of their range the following qualifying condition

$$0 \leqslant f'_{1hk}(x_{1h}) \leqslant q_{1hk} \leqslant f'_{2hk}(x_{2h}) \leqslant q_{2hk} \leqslant \ldots \leqslant f'_{ihk}(x_{ih}) \leqslant$$

$$\leqslant q_{ihk} \leqslant \ldots \leqslant f'_n(x_{nh}) , \qquad (29)$$

then the optimal policy (25), solving the problem in the linear case, is also optimal in the non-linear case.

One should just remark that the qualifying condition (29) can turn out to be rather restrictive, since in fact it requires that the maximum of the marginal conversion cost in a country is not larger than the minimum of the same cost in the next country. This condition is indeed largely fulfilled in the linear case. We must then demonstrate that, in the hypothesis (29), the policy described by (25) is optimal also in the non-linear case (20). Also, that the residual resources, in units of h or k respectively in (26) and (27), give now

$$Y_k = \sum_{i=s+2}^{n} R_{ik} + f_{s+1,hk}\left(\sum_{i=1}^{s+1} R_{ih} - X_h\right), \quad (X_h \in I_{s+1,h}) \; . \qquad (30)$$

In the linear case, this coincides with (26). Also, that it is the value of the objective function of the problem (20) corresponding to the policy (25). For our purpose it is sufficient to prove that by modifying the policy (25) in any of the following ways:

$$x_{1h} = R_{1h};\; x_{2h} = R_{2h};\ldots;\; x_{jh} = R_{jh} - \sigma;\ldots;\; x_{sh} = R_{sh};$$

$$x_{s+1,h} = X_h - \sum_{i=1}^{s} R_{ih} - \Delta,\; x_{s+2,h} = 0;\ldots;\; x_{th} = \sigma;\ldots;$$

$$x_{uh} = \Delta\; ;\ldots;\; x_{nh} = 0; \quad (j = 1, \ldots, s;\; t = s+2, \ldots, n;$$

$$u = s+2, \ldots, n) \qquad (31)$$

(or similar ways), with σ and Δ chosen in order to satisfy the constraints of the problem (20), the corresponding value of the objective function is less than shown in (30). Or that (referring to the variation rules in (31), with similar procedures for other variations of the policy) the inequality holds

$$f_{jhk}(\sigma) + f_{s+1,hk}\left(\sum_{i=1}^{s+1} R_{ih} - X_h + \Delta\right) + \sum_{i=s+2}^{n} R_{ik} - R_{tk} - R_{uk} +$$

$$+ f_{thk}(R_{th} - \sigma) + f_{uth}(R_{uh} - \Delta) \leqslant \sum_{i=s+2}^{n} R_{ik} +$$

$$+ f_{s+1,hk}\left(\sum_{i=1}^{s+1} R_{ih} - X_h\right) \quad ; \qquad (32)$$

this, according to (12) and (19), is equivalent to

$$\left\{f_{jhk}(\sigma) - f_{jhk}(0)\right\} - \left\{f_{thk}(R_{th}) - f_{thk}(R_{th} - \sigma)\right\} +$$

$$+ \left\{f_{s+1,hk}\left(\sum_{i=1}^{s+1} R_{ih} - X_h + \Delta\right) - f_{s+1,hk}\left(\sum_{i=1}^{s+1} R_{ih} - X_h\right)\right\} -$$

$$- \left\{f_{uhk}(R_{uh}) - f_{uhk}(R_{uh} - \Delta)\right\} \leqslant 0 \ , \qquad (33)$$

hence also to

$$\left\{f'_{jhk}(\xi_j) - f'_{thk}(\xi_t)\right\} \sigma + \left\{f'_{s+1,hk}(\xi_{s+1}) - f'_{uhk}(\xi_u)\right\} \Delta \leqslant 0, \qquad (34)$$

ξ_j, ξ_t, ξ_{s+1}, ξ_u being conveniently chosen according to Lagrange formula. But (34) holds since σ and Δ are non negative and each expression enclosed by braces is non positive according to the qualifying condition (29) and since j precedes t and s+1 precedes u.

8. THE LAW OF RECIPROCITY.

One should remind that two problems of conditioned extremum are reciprocal of each other when the function for what the extremum is sought in one of them, appears in the other one, made equal to a constant, as a constraint, while all other possible constraints are the same in the problems [6]. E.g., the problem

$$P_M(\xi) : \begin{cases} \max f(X) \overset{\text{def}}{===} F_M(\xi) , \\ g(X) = \xi \\ X \in D \end{cases}$$

where D is an assigned set in the space where X is any element, admits the following couple of reciprocal problems:

$$Q_m(\eta) : \begin{cases} \min g(X) \overset{\text{def}}{===} G_m(\eta) \\ f(X) = \eta \\ X \in D \end{cases} ;$$

$$Q_M(\eta) : \begin{cases} \max g(X) \overset{\text{def}}{===} G_M(\eta) \\ f(X) = \eta \\ X \in D \end{cases} ;$$

for such problems the following result holds, as recently pointed out by F. Giannessi:

- if X° is a maximum for the problem $P_M(\xi)$ and if $F_M(\xi)$ is decreasing, then X° is a maximum for the problem $Q_M(F_M(\xi))$. Moreover, the equality holds

  $$G_M(F_M(\xi)) = \xi \quad ,$$

  characterizing $G_M(\eta)$ as the inverse function of $F_M(\xi)$.

Applying it to our case, we find that the two problems

$$\begin{cases} \max \sum_{i=1}^{n} f_{ihk}(R_{ih} - x_{ih}) \overset{def}{===} Y_k(X_h) \,, \\ \sum_{i=1}^{n} x_{ih} = X_h \,, \quad (X_h \in H) \,, \\ 0 \leqslant x_{ih} \leqslant R_{ih} \,, \quad (i = 1,\dots, n) \end{cases} \tag{35}$$

$$\begin{cases} \max \sum_{i=1}^{n} x_{ih} \overset{def}{===} X_h(Y_k) \,, \\ \sum_{i=1}^{n} f_{ihk}(R_{ih} - x_{ih}) = Y_k \,, \quad (Y_k \in K) \,, \\ 0 \leqslant x_{ih} \leqslant R_{ih} \,, \quad (i = 1,\dots, n) \end{cases} \tag{36}$$

are reciprocal of each other according to the above analytical definition. Having in mind the economical meaning of the first problem, the one of the second is sought. In it, one can consider as unknown

$$f_{ihk}(R_{ih} - x_{ih}) \overset{def}{===} y_{ik} \,, \quad (i = 1,\dots, n) \,, \tag{37}$$

and look at it as an amount of resource (in standard units) that the country P_i could convert to the production of an assigned amount Y_k (of the standard good k), so that the whole production of each country satisfies the equality

$$\sum_{i=1}^{n} f_{ihk}(R_{ih} - x_{ih}) = \sum_{i=1}^{n} y_{ik} = Y_k \,. \tag{38}$$

From this point of view, the amount $x_{ih} = R_{ih} - (R_{ih} - x_{ih})$ represents the residual resources of the country P_i (in units of h) to be fully

exploited in the production of h so that the maximum community yield is obtained there, and it is given by the addition $\sum_{i=1}^{n} x_{ih}$.
But the residual resources of the country P_i are then given, in standard units, by $R_{ik} - y_{ik}$, since, in P_i, the amount y_{ik} in standard units, represents the resources already involved in the production of the assigned amount Y_k of the standard good k. According to the equivalence expressed in (12), the inequality holds

$$f_{ikh}(R_{ik} - y_{ik}) = x_{ih} , \tag{39}$$

so that the problem (36), with the change of variables described in (37) becomes the equivalent problem

$$\begin{cases} \max \sum_{i=1}^{n} f_{ikh}(R_{ik} - y_{ik}) \overset{\text{def}}{\equiv} X_h(Y_k) \\ \sum_{i=1}^{n} y_{ik} = Y_k \ , \ (Y_k \in K) \ , \\ 0 \leqslant y_{ik} \leqslant R_{ik}, \ (i = 1, \dots, n) \ . \end{cases} \tag{40}$$

The structure of this problem is analitically the same as the one of the problem (35). From the point of view of production, it refers to the following reciprocal problem:

- the exploitation of the resources required to produce the total amount Y_k of the standard good k should be partitioned among the partners of the community in such a way that the possible residual resources in the various countries, if fully used to produce the good h, give the maximum yield.

Now, for the problem (35), the function $Y_k = Y_k(X_h)$, giving the maximum, is, as shown in the previous section, the one given in (30) for $X_h \in I_{s+1,h}$. By fully writing the defining relation one has

$$Y_k = Y_k(X_h) = \begin{cases} \sum_{i=2}^{n} R_{ik} + f_{ihk}(R_{ih} - X_h), & \text{per } X_h \in I_{1,h} \\ \sum_{i=3}^{n} R_{ik} + f_{2hk}\left(\sum_{i=1}^{2} R_{ih} - X_h\right), & \text{per } X_h \in I_{2,h} \\ \dots\dots\dots\dots\dots\dots\dots\dots \\ \sum_{i=s+2}^{n} R_{ik} + f_{s+1,hk}\left(\sum_{i=1}^{s+1} R_{ih} - X_h\right), & \text{per } X_h \in I_{s+1,h} \\ \dots\dots\dots\dots\dots\dots\dots\dots \\ R_{nk} + f_{n-1,hk}\left(\sum_{i=1}^{n-1} R_{ih} - X_h\right), & \text{per } X_h \in I_{n-1,h} \\ f_{nhk}\left(\sum_{i=1}^{n} R_{ih} - X_h\right), & \text{per } X_h \in I_{nh} \end{cases} \tag{41}$$

This is sufficient to show the decrease after the qualifying condition (29). Then, according to the result by Giannessi, the two problems (35) and (36) have the same point of maximum as soon as $Y_k = Y_k(X_h)$ is inserted in (36). Moreover, the function $X_h(Y_k)$, giving the maximum in the problem (36) and also in (40), is the inverse of $Y_k(X_h)$. After (41), the definition of $X_h(Y_k)$ is then the following

$$X_h = X_h(Y_k) = \begin{cases} R_{1h} - f_{1kh}(Y_k - \sum_{i=2}^{n} R_{ik}) , & Y_k \in J_{1k} , \\ \sum_{i=1}^{2} R_{ih} - f_{2kh}(Y_k - \sum_{i=3}^{n} R_{ik}) , & Y_k \in J_{2k} , \\ \dots\dots\dots\dots\dots\dots & \\ \sum_{i=1}^{s+1} R_{ih} - f_{s+1,kh}(Y_k - \sum_{i=s+2}^{n} R_{ik}), & Y_k \in I_{s+1,k}, \\ \dots\dots\dots\dots\dots\dots & \\ \sum_{i=1}^{n-1} R_{ih} - f_{n-1,kh}(Y_k - R_{nk}) , & Y_k \in J_{n-1,k} , \\ \sum_{i=1}^{n} R_{ih} - f_{nkh}(Y_k) , & Y_k \in J_{nk} , \end{cases} \tag{42}$$

and the optimal policy to achieve the maximum in the problem (40) stemming out of it, is the corresponding of (25) in the change of variables defined in (37). So

$$y_{1k} = 0;\ y_{2k} = 0;\dots;\ y_{sk} = 0;\ y_{s+1,k} = f_{s+1,hk}\left(\sum_{i=1}^{s+1} R_{ih} - X_h\right)$$

$$y_{s+2} = R_{s+2,k};\ \dots\ ;\ y_{nk} = R_{nk}\ . \tag{43}$$

This, in the linear case, identifies with (28).

Then, the <u>law of reciprocity</u> becomes, in our case, the following statement:

- problem (20) and problem (40) are reciprocal; the functions $Y_k = Y_k(X_h)$ and $X_h = X_h(Y_k)$, giving the two maxima, are the inverse of each other. The optimal policy of one of them is given by the residual resources that the optimal policy of the other one has left in the various countries.

9. THE FRONTIER OF THE MAXIMUM EFFICIENCY. THE CURVE OF THE COMPARED COSTS.

In the classical theory the function $Y_k = Y_k(X_h)$, or more precisely its geometrical representation, is called the frontier of maximum efficiency. It is understood to be the maximum efficiency of the community in producing the amount X_h of h. The same holds for the (inverse) function $X_h = X_h(Y_k)$. Considering this meaning, after the criterion adopted in § 2 to determine the global community cost of a good, we can then conclude that the community cost of any amount X_h of h (and similarly for Y_k of k) is given by the difference

$$Z_k = R_k - Y_k(X_h) \ , \ (X_h \in H) \ . \tag{44}$$

Using the symbols introduced in § 3, considering (41), it is defined as follows

$$Z_k = \begin{cases} 0 & , \text{ per } X_h = \bar{H}_0 \\ R_{1k} - f_{1hk}(R_{1h} - X_h) & , \text{ per } X_h \in H_1 \\ R_{1k} & , \text{ per } X_h \in \bar{H}_1 \\ \dots\dots\dots\dots\dots\dots\dots\dots & \\ \sum_{i=1}^{s} R_{ik} & , \text{ per } X_h = \bar{H}_s \\ \sum_{i=1}^{s+1} R_{ik} - f_{s+1,hk}\left(\sum_{i=1}^{s+1} R_{ih} - X_h\right) & , \text{ per } X_h \in H_{s+1} \\ \dots\dots\dots\dots\dots\dots\dots\dots & \\ \sum_{i=1}^{n} R_{ik} - f_{nhk}\left(\sum_{i=1}^{n} R_{ih} - X_h\right) & , \text{ per } X_h \in H_n \\ \sum_{i=1}^{n} R_{ik} = R_k & , \text{ per } X_h = \bar{H}_n \ . \end{cases} \tag{45}$$

Since Z_k is the complement (with respect to R_k) of $Y_k(X_h)$, it is evident that the geometric representation of the latter yields the representation of Z_k. We then consider the geometric representation of $Y_k = Y_k(X_h)$. The classical theory calls it properly the frontier of maximum efficiency. After (41), this representation is a decreasing curve - composed polygon, i.e. a continuous decreasing curve, composed by several adjacent arcs, related to the ranges $H_{s,h}$, (s = 1,..., n). Clearly, the arcs become parts of straight lines in the linear case. As an example, in the linear case and for a community of four countries, the frontier of maximum efficiency may look like in the following figure

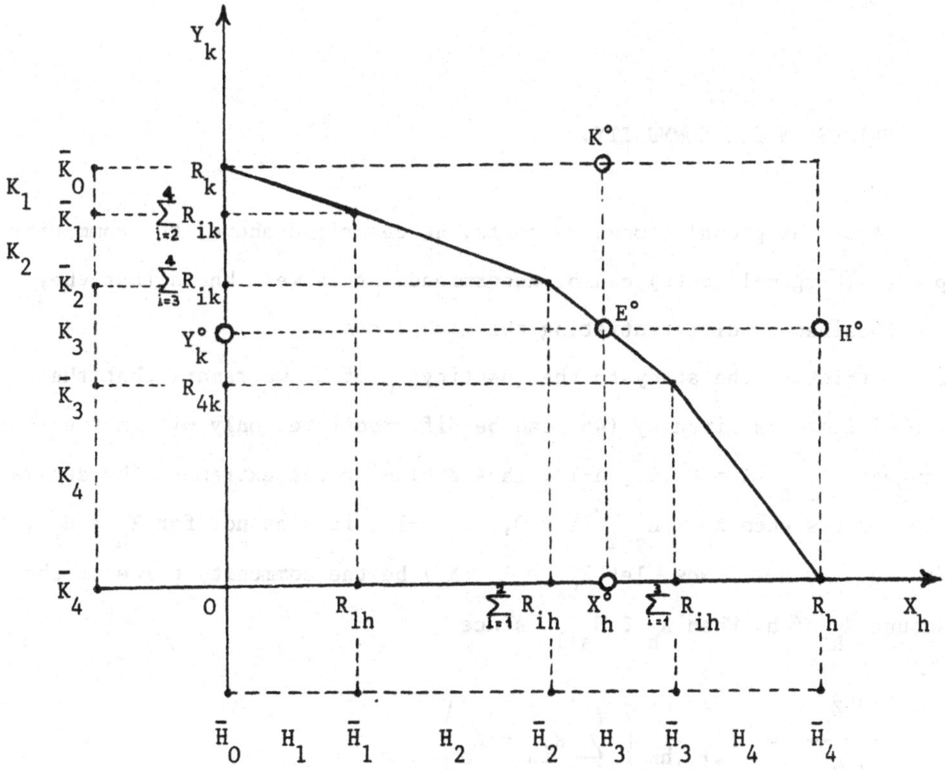

One can see that, choosing any X°_h in $(0, R_h)$, the corresponding Y°_k of the point E° gives the maximum efficiency of the community in the production of the standard good k under the condition of producing the quantity X°_h of h.

It follows that the lenght of the segment $E^\circ K^\circ$ (being the complement of Y°_k to reach R_k) represents the global community cost (in standard units of k) of the amount X°_h of h.

Similar remarks hold if in $(0, R_k)$ the amount Y°_k of k is established. Its global community cost (in h units) is then given by the lenght of $E^\circ H^\circ$.

For the above facts, the frontier of maximum efficiency is also called, in the classical theory, the curve of the compared costs.

10. PRICES IN THE COMMUNITY.

From the global community costs, as described above, the community prices (marginal costs) can be determined. It is well known that they are obtained by differentiating the global costs.

By restricting the study to the quantity X_h of h, we remark that the global cost, as given by (45) can be differentiated only within the ranges $I_{s+1,h}$ $(s = 0,\dots, n-1)$, thus excluding the extremes. The derivative exists when $X_h \in H_{s+1}$ $(s = 0,\dots, n-1)$, it does not for $X_h = \bar{H}_t$, $(t = 0,\dots, n-1)$. Now, let $\lambda_{hk} = \lambda_{hk}(X_h)$ be the community price of the amount X_h of h. When $X_h \in H_{s+1}$, since

$$\frac{d Z_k}{d X_h} = f'_{s+1,hk}\left(\sum_{i=1}^{s+1} R_{ih} - X_h\right) ,$$

is the community price, λ_{hk} ranges in

$$\bar{\Lambda}_{s+1,h} \overset{\text{def}}{\equiv} \min_{x_{s+1,h} \in (0,R_{s+1,h})} f'_{s+1,hk}(x_{s+1,h}) \leq \lambda_h \leq$$

$$\leq \max_{x_{s+1,h} \in (0,R_{s+1,h})} f'_{s+1,hk}(x_{s+1,h}) \; .$$

In the linear case, this degenerates in the point $\lambda_{hk} = q_{s+1,hk}$. Instead, when $X_h = \bar{H}_t$, $(t = 0,\ldots, n-1)$, the global cost cannot be differentiated: the optimal community policy does not expect in any country the simultaneous production of the two goods. Precisely, if $X_h = \bar{H}_{s+1}$, the optimal policy of the community is such that the first s countries (following the order that qualifies them to produce the good h) produce only the good h, while the other ones (more qualified to produce k) produce only the good k. A point of the curve of maximum efficiency, corresponding to such a critical value of X_h, is called in the classical theory <u>Ricardo point</u>. When X_h is in the abscissa of a Ricardo point, the derivate of the global cost does not exist. Anyway, we define the community price also in such points, by conventionally identifying it with any number λ_{hk} belonging respectively to the intervals

$$\Lambda_0 \overset{\text{def}}{\equiv} 0 \leq \lambda_{hk} \min_{x_{1h} \in (0,R_{1h})} f'_{1hk}(x_{1h}) \; , \qquad \text{if } X_h = \bar{H}_0 \; ,$$

$$\Lambda_{s+1} \overset{def}{\equiv} \max_{x_{sh}\in(0,R_{sh})} f'_{shk}(x_{sh}) \leq \lambda_{hk} \leq$$

$$\leq \min_{x_{s+1,h}\in(0,R_{s+1,h})} f'_{s+1,hk}(x_{s+1,h}) \quad , \qquad \text{if } X_h = \bar{H}_{s+1} \ ,$$

$$(s = 0, \ldots, n-2) \ ,$$

$$\Lambda_n \overset{def}{\equiv} \max_{x_{nh}\in(0,R_{nh})} f'_{nhk}(x_{nh}) \leq \lambda_{hk} < +\infty \quad , \quad \text{if } X_h = \bar{H}_n \ .$$

This way the community price of the quantity X_h of h is defined for each $X_h \in (0, R_h)$. Its determination can be coded with the following rule:

- when X_h does not coincide with the abscissa of any Ricardo point, the community price λ_{hk} (of X_h) varies between the minimum and the maximum local price of the country to be selected from what follows. It should be the last one, in the rank qualifying the countries for the production of h, to be called by the optimal community policy to help in the production of the fixed quantity X_h of h.

  Otherwise, when X_h is identified with the abscissa $\bar{H}_0$, $\bar{H}_1$, ..., $\bar{H}_n$ of a Ricardo point, the community price is any number λ_h belonging to the intervals Λ_0, Λ_1, ..., Λ_n.

As a conclusion we can remark that by increasing the quantity X_h (of h) which should be produced by the community, the optimal policy for the community requires that the countries where the local price is higher, which are then, in order, the ones less qualified to produce the good h, contribute to the production of h. Since, according to the previous rule the local price (the one of the last producing country) determines the price in the community, it follows that the price in the community increases with X_h. But at the same time the local price of the good h in the countries more qualified to produce it tends to lower. In fact, since those countries have to produce only the good h (and nothing of the good k), the gap between their capability to produce h and their bad qualification to produce k is enhanced. In those countries then, the conversion h → k becomes more difficult and then more expensive. This way, the local price of the standard product k is increasing and the one of the good h is decreasing. Another remark which is worth mentioning concerns the local production policy that each country would choose according to their own advantage (without any regard for the community), if the only information at hand were the level of the price in the community. Even ignoring the optimal policy (in the community sense) for sure they would locally adapt such a policy. Indeed, knowing that the community price is, for example, ranging in $\Lambda_{s+1,hk}$, the countries P_i ($i \leq s$) would be compelled to produce only the good h since in these countries the community price is higher than the local one. Things are just opposite where $i > s$.

Concluding, a production policy which is optimal for the community is also locally optimal.

The above conclusion emphasizes that a policy which is locally optimal for economic purpose (i.e., it yields the maximum local returns, according to the community prices) identifies with the directions of a policy which is optimal for the community production (i.e., it yields the maximum producible amount of either of two goods when the quantity of the other one is fixed).

The question then arises whether the community optimal policy for the production quantity is community optimal as well for the economic purposes, using the community prices as standard. The answer is definitely affirmative. Indeed, the community optimal policy for quantitative purposes, giving the maximum local returns in each country, implies global community returns which, being the sum of the local maxima, cannot be less than with any other policy.

REFERENCES

1. Volpato, M., *Studi e Modelli di Ricerca Operativa,* U.T.E.T., Torino, 1971, chap. 1, section 2, lemmas 3 - 4, pp. 34-35; chap. 2, p. 285.
2. Cucconi, O., Sopra un particolare problema (non lineare) di distribuzione, in Volpato [1], chap. 3.
3. Wilkinson, C. and Gupta, S.K., Allocating Promotional Effort to Competing Activities: A Dynamic Approach, in *Proceedings of the Vth International Conference on Operational Research*, Venice, 1969, J. Lawrence ed., Tavistock Publications, London, 1970.
4. Castellani, G., *Su un particolare problema di teoria della ricerca,* Rendiconti del Comitato per il potenziamento in Venezia degli studi economici, CEDAM, Padova, 1971.
5. Volpato, M., *Studi e Modelli di Ricerca Operativa*, U.T.E.T., Torino, 1971, chap. 2, section 2, p. 277.
6. Giannessi, F., Sulla legge di reciprocità nei problemi di massimo e minimo condizionati, in Volpato [1], chap. 7, section 2, p. 1053.

EXPLICIT SOLUTION FOR A CLASS OF ALLOCATION PROBLEMS (°)

Giovanni Castellani
Department of Mathematics
University of Venice

1. INTRODUCTION.

In this paper, we deal with the classical allocation problem of optimizing the function

$$\sum_{i=1}^{n} f_i(x_i) \tag{1}$$

under the following constraints

$$\begin{cases} \sum_{i=1}^{n} x_i \leqslant C & (C > 0) \\ x_i \geqslant 0 & (i = 1, 2, \dots, n) \end{cases} \tag{2}$$

(°) Partially supported by the C.N.R. Research Group of Functional Analysis and Applications.

that can be solved by applying the dynamic programming algorithm [1]. Under particular assumptions on $f_i(x_i)$, the explicit solution can be obtained for the problem above [2].

Aim of the paper is to generalize this result and the explicit solution is determined when the functions $f_i(x_i)$ are differentiable and strictly convex or concave (according to the nature of the problem). Moreover, the dynamic programming algorithm will be simplified by giving a criterion, which shows whether we need examine the following steps to find the solution.

There are two different approaches to do this: the former uses the Kuhn-Tucker theorem (or the Gibbs's lemma) [3], the latter is related to the dynamic programming method.

For both of these cases, the sequence $f_i'(0)$ is assumed to be monotonic (nonincreasing or nondecreasing according to the nature of the problem), so that the positive components preceed all the zero components in the optimizing vector. This will make it easier to achieve the goal.

2. STATEMENT OF THE PROBLEM.

Let us consider the following problem

$$\max \sum_{i=1}^{n} f_i(x_i) \tag{3}$$

$$\begin{cases} \sum_{i=1}^{n} x_i \leqslant C \\ x_i \geqslant 0 \end{cases}$$

where $f_i(x_i)$, for i = ..., n, is a twice-differentiable and strictly concave function in $[0, C]$.

Under these assumptions, problem (3) has only one solution.

As the inequality

$$f_i'(0) \leqslant 0 \tag{4}$$

implies that $f_i(x_i)$ is a decreasing function in $[0, C]$ and the corresponding i-th component in the maximizing vector is zero, we can restrict ourselves to the problem:

$$\max \sum_{i=1}^{m} f_i(x_i) \tag{5}$$

$$\begin{cases} \sum_{i=1}^{m} x_i \leqslant C \\ x_i \geqslant 0 \end{cases}$$

with

$$f_i'(0) > 0 \quad , \quad (i = 1, 2, \dots, m) \tag{6}$$

and $m \leqslant n$.

Moreover, being the functions $f_i'(x_i)$ decreasing in $[0, C]$, the equation

$$f_i'(x_i) = 0 \tag{7}$$

has one real root $\bar{x}_i$, at most, in $[0, C]$ with $\bar{x}_i \neq 0$ for assumption (6). If such a root exists for any $i = 1, 2, \dots, m$ and satisfies the inequality

$$\sum_{i=1}^{m} \bar{x}_i \leqslant C , \qquad (8)$$

then the point

$$\bar{x} = (\bar{x}_1, \bar{x}_2, \ldots, \bar{x}_m)$$

solves problem (5).
Otherwise, if the maximizing point is

$$x^* = (x_1^*, x_2^*, \ldots, x_m^*) , \qquad (9)$$

the following relationship holds

$$x_i^* < \bar{x}_i , \qquad (10)$$

for any index i, such that (7) has a root and the first constraint is fulfilled

$$\sum_{i=1}^{m} x_i^* = C . \qquad (11)$$

(11) is guaranteed [4] because, if $\bar{x}_i$ exists, $f_i(x_i)$ increases in $\left[0, \bar{x}_i\right]$; alternatively, if not, it increases in $[0, C]$.
Moreover, inequality (10) can be obtained by applying the Kuhn - Tucker theorem.
Indeed, when the functions $f_i(x_i)$ are strictly concave, (9) is a solution for problem (5), if and only if there is a λ^* multiplier which makes system (12) consistent,

$$\begin{cases} f_i'(x_i^*) \leqslant \lambda^* & (i = 1, 2, \ldots, m) \quad (12.1) \\ x_i^* \geqslant 0 & (i = 1, 2, \ldots, m) \quad (12.2) \\ x_i^* f_i'(x_i^*) = x_i^* \lambda^* & (i = 1, 2, \ldots, m) \quad (12.3) \\ \lambda^* \geqslant 0 & \quad (12.4) \\ \sum_{i=1}^{m} x_i^* \leqslant C & \quad (12.5) \\ \lambda^* \left(\sum_{i=1}^{m} x_i^* - C \right) = 0 & \quad (12.6) \end{cases}$$

We shall now prove that, under our assumptions, it is $\lambda^* > 0$.
Indeed, if we assume $\lambda^* = 0$, for (12.1) the equality $x_i^* = 0$ implies $f_i'(0) \leqslant 0$, that contradicts (6) and we must take $x_i^* > 0$ for any index i. Then, from (12.3) it follows $f_i'(x_i^*) = 0$ for i = 1, 2, ..., n, and this cannot hold, if not all the equations (7) have a root in [0, C].
On the other hand, if such a root exists for any index i, but inequality (8) does not hold, then it results $x_i^* = \bar{x}_i$ (i = 1, 2, ..., m) and (8) follows from (12.5), contradicting our assumptions.
Thus the proof is complete; moreover, we can see that from (12.6) and $\lambda^* > 0$, we find that (11) holds.
Now we can prove inequality (10) for the values of i such that (7) has a root $\bar{x}_i \in [0, C]$.
Indeed, if the corresponding x_i^* is positive, (10) follows from (12.3) and from the fact that $f_i'(x_i)$ is decreasing and $f_i'(\bar{x}_i) = 0$, $\lambda^* > 0$; on the other hand, if $x_i^* = 0$, inequality (10) is guaranteed by (6).
Finally, when (8) does not hold, problem (5) has the same solution as

$$\operatorname{Max} \sum_{i=1}^{m} f_i(x_i) \tag{13}$$

$$\begin{cases} \sum_{i=1}^{m} x_i = C \\ x_i \geqslant 0 \end{cases}$$

which we shall be concerned with, in the following chapters.

3. A PROPERTY OF THE MAXIMIZING VECTOR FOR PROBLEM (13).

Let us consider problem (13) and its maximizing vector (9) : it is easy to prove that for a pair of indexes i and j with i < j (i, j = = 1, 2, ..., m), when we set

$$x_i^* + x_j^* = z^* \quad ,$$

(x_i^*, x_j^*) is the only solution for the problem

$$\max \left[f_i(x_i) + f_j(x_j) \right] \tag{14}$$

$$\begin{cases} x_i + x_j = z^* \\ x_i, x_j \geqslant 0 \quad . \end{cases}$$

Indeed, let us assume a different solution $(\hat{x}_i, \hat{x}_j)$ for problem (14): then we can construct a new vector

$$(x_1^*, x_2^*, \ldots, x_{i-1}^*, \hat{x}_i, x_{i+1}^*, \ldots, x_{j-1}^*, \hat{x}_j, x_{j+1}^*, \ldots, x_m^*) \; ,$$

that satisfies the constraint $\sum_{i=1}^{m} x_i = C$, because of the equality $x_i^* + x_j^* = \hat{x}_i + \hat{x}_j$, and makes the objective function $\sum_{i=1}^{m} f_i(x_i)$ greater that the value obtained at point (9).

That being stated, we rewrite problem (14) in the following way:

$$\max_{0 \leqslant x_j \leqslant z^*} \left\{ f_i(z^* - x_j) + f_j(x_j) \right\} . \tag{15}$$

Let us set

$$g(x_j) = f_i(z^* - x_j) + f_j(x_j) ,$$

so that we have

$$g'(x_j) = -f_i'(z^* - x_j) + f_j'(x_j) .$$

We are now going to prove the following

<u>Lemma 1</u>.

Let us assume

$$f_i'(0) \geqslant f_j'(0) :$$

in correspondence to one real number α*, at most, with* $0 \leqslant \alpha < z^*$, *the following inequalities hold*

$$g'(x_j) > 0 \qquad \textit{for} \;\; 0 \leqslant x_j < \alpha \;, \textit{if} \; \alpha > 0 \; ; \tag{16}$$

$$g'(x_j) = 0 \qquad \textit{for} \;\; x_j = \alpha \tag{17}$$

$$g'(x_j) < 0 \quad \textit{for either } \alpha < x_j \leqslant z^* \textit{ or } \; 0 \leqslant x_j \leqslant z^* \tag{18}$$
$$\textit{if } \alpha \textit{ does not exist.}$$

Proof.

As $f_i(x_i)$ and $f_j(x_j)$ are strictly concave, $f'_j(x_j)$ is decreasing with respect to x_j and $f'_i(z^* - x_j)$ is increasing with respect to x_j in $[0, C]$. Then we can find one value, at most, $x_j = \alpha$, such that

$$f'_i(z^* - x_j) = f'_j(x_j) \; ,$$

or, equivalently,

$$g'(x_j) = 0 \; . \tag{19}$$

Moreover, we see that it cannot be $\alpha = z^*$, as the relation

$$g'(z^*) = -f'_i(0) + f'_j(z^*) < 0 \tag{20}$$

follows from the inequalities

$$f'_i(0) \geqslant f'_j(0) > f'_j(z^*) \; ,$$

the former of which holds for our assumptions and the latter is true, because $f'_j(x_j)$ decreases in $[0, z^*]$.

We have proved that (17) holds.

Finally, as $g'(x_j)$ is a continuous function, and the only root α of (19) is simple, we get (18) and (16).

Lemma 2.

Let us assume the following inequalities hold

$$f'_1(0) \geq f'_2(0) \geq \dots \geq f'_m(0) \ . \tag{21}$$

Then the maximizing vector $x^* = (x^*_1, x^*_2, \dots, x^*_m)$ *for problem* (13) *satisfies the inequalities*

$$| f''_1(\delta_1 x^*_1) | x^*_1 \geq |f''_2(\delta_2 x^*_2)| \ x^*_2 \geq \dots \geq |f''_m(\delta_m x^*_m)| \ x^*_m \tag{22}$$

with $0 < \delta_i < 1$.

<u>Proof.</u>

If the root α, defined in Lemma 1, does not exist, the solution to (15) is

$$x^*_j = 0 \ ,$$

as we have $g'(x_j) < 0$ over the whole closed interval $\left[0, z^*\right]$.
In this case, the solution to the equivalent problem (14) is

$$x^*_i = z^* \qquad x^*_j = 0 \ ,$$

which satisfies the inequality

$$|f''_i(\delta_i x^*_i)| \ x^*_i \geq |f''_j(\delta_j x^*_j)| \ x^*_j \ .$$

If the root α does not exist, the solution of (15) is

$$x^*_j = \alpha \ ,$$

so that the equivalent problem (14) is solved by

$$x_i^* = z^* - \alpha \,, \quad x_j^* = \alpha$$

and we get

$$f_i'(x_i^*) = f_j'(x_j^*) \,. \tag{23}$$

As we can write

$$f_i'(x_i^*) = f_i'(0) + x_i^* \, f_i''(\delta_i \, x_i^*) \qquad 0 < \delta_i < 1$$

and

$$f_j'(x_j^*) = f_j'(0) + x_j^* \, f_j''(\delta_j \, x_j^*) \qquad 0 < \delta_j < 1 \,,$$

from (23) and from

$$f_i'(0) \geqslant f_j'(0) \,,$$

it results

$$x_i^* \, f_i''(\delta_i \, x_i^*) \leqslant x_j^* \, f_j''(\delta_j \, x_j^*) \,,$$

or, equivalently, from the strict concavity of $f_i(x_i)$ and $f_j(x_j)$, we have

$$|f_i''(\delta_i \, x_i^*)| \, x_i^* \geqslant |f_j''(\delta_j \, x_j^*)| \, x_j^* \,. \tag{24}$$

The proof of Lemma 2 is now complete, because (24) holds for any pair of indexes i, j, with i < j.

Thus, we can state the following property for the maximizing vector of (13):

Let $f_i'(0)$ *be ordered according to* (21), *then if the j-th component of the vector, which maximizes* (13) *is positive, all the preceeding components are positive; if it is zero, all the following ones are zero.*

4. SOLUTION OF PROBLEM (13) BY MEANS OF THE KUHN - TUCKER THEOREM.

For our particular problem (13) with strictly concave functions $f_i(x_i)$, the Kuhn - Tucker theorem, we mentioned above, states that the vector $x^* = (x_1^*, x_2^*, \ldots, x_m^*)$ solves problem (13), if and only if it exists a λ^* multiplier such that

$$\begin{cases} f_i'(x_i^*) \leqslant \lambda^* \qquad (i = 1, 2, \ldots, m) & (25.1) \\ x_i^* \geqslant 0 & (25.2) \\ x_i^* \, f_i'(x_i^*) = x_i^* \, \lambda^* & (25.3) \\ \sum_{i=1}^{m} x_i^* = C & (25.4) \end{cases}$$

We remark that, under our particular assumptions, the same result has been stated by J. Willard Gibbs and from it we can derive the following property.

Let us assume the functions $f_i(x_i)$ *satisfy inequality* (21) *and let* r *be the greatest index* i, i = 1, 2, ..., m, *such that the root* λ_r *of the equation*

$$\sum_{i=1}^{r} f_i'^{-1}(\lambda) = C \tag{26}$$

satisfies the relationship

$$f_r'(0) > \lambda_r . \tag{27}$$

Then the maximizing vector for problem (13) *is given by*

$$\begin{cases} x_i^* = f_i'^{-1}(\lambda_r) & (i = 1, 2, \ldots, r) \\ x_i^* = 0 & (i = r+1, r+2, \ldots, m) \end{cases} \tag{28}$$

and the maximum value is

$$\sum_{i=1}^{r} f_i(f_i'^{-1}(\lambda_r)) + \sum_{i=r+1}^{m} f_i(0) .$$

First of all, we remark that, as $f_i(x_i)$ is a strictly concave function, its derivative $f_i'(x_i)$ decreases and can be inverted, so that the sum $\sum_{i=1}^{r} f_i'^{-1}(\lambda)$ is decreasing on the set $\bigcap_{i=1}^{r} \left[f_i'(C), f_i'(0) \right]$ and equation (26) has one real root, at most.

In the particular case r = 1, equation (26) has a real root satisfying (27): indeed, from

$$f_1'^{-1}(\lambda) = C ,$$

and, consequently

$$\lambda_1 = f_1'(C) ,$$

it results

$$f_1'(0) > f_1'(C) ,$$

as $f_1'(x_1)$ is a decreasing function.

We are now going to prove that (28) maximizes problem (13).

As (21) and (27) imply

$$f_i'(0) > \lambda_r \qquad (i = 1, 2, \ldots, r) , \tag{29}$$

assuming $\lambda^* = \lambda_r$, we cannot find $x_i^* = 0$ for (25.1). Thus, as we deduce $x_i^* > 0$ from (25.2), we have

$$x_i^* = f_i'^{-1}(\lambda_r) \qquad (i = 1, 2, \ldots, r) ,$$

that follows from (25.3).

Moreover, the inequality $x_{r+1}^* > 0$ contradicts the assumption that r is the greatest index satisfying (30).

Indeed, if this is true, from (25.3) - (25.4) it follows

$$x_i^* = f_i'^{-1}(\lambda_{r+1}) \qquad (i = 1, 2, \ldots, r+1)$$

and, consequently,

$$f'_{r+1}(x^*_{r+1}) = \lambda_{r+1} \quad :$$

but the inequality

$$f'_{r+1}(0) > \lambda_{r+1}$$

contradicts our assumptions.

Now, taking into account that in the maximizing vector the positive components preceed the zero components, we can conclude that (28) solves (13).

5. SOME APPLICATIONS.

In this section, we shall solve some particular problems, by applying the rule described above. Further examples can be found in Giannessi[5, 6].

a) Let us consider the problem[7]

$$\max \sum_{i=1}^{m} (a_i + b_i x_i + c_i x_i^2) \tag{30}$$

$$\begin{cases} \sum_{i=1}^{m} x_i \leqslant C \\ x_i \geqslant 0 \end{cases}$$

with $C > 0$ and $c_i < 0$.

If some b_i are nonpositive, the corresponding components x_i^* of the maximizing vector are zero: therefore we can assume

$$b_i > 0 \qquad (i = 1, 2, \dots, m)$$

and the relationships

$$b_1 \geqslant b_2 \geqslant \dots \geqslant b_m$$

imply inequalities (21).

For coefficients linked by the inequality

$$\sum_{i=1}^{m} - \frac{b_i}{2c_i} \leqslant C ,$$

the solution of (30) is given by

$$x_i = - \frac{b_i}{2c_i} \qquad (i = 1, 2, \ldots, m) ,$$

and the maximum value for the objective function becomes

$$\sum_{i=1}^{m} \left(a_i - \frac{b_i}{4c_i} \right) .$$

Alternatively, let us assume the inequality

$$\sum_{i=1}^{m} - \frac{b_i}{2c_i} > C$$

holds.

If r is the greatest index i, i = 1, 2, ..., m, such that the equation

$$\sum_{i=1}^{r} \frac{\lambda - b_i}{2c_i} = C \tag{31}$$

has a root λ_r, satisfying the inequality

$$b_r > \lambda_r , \tag{32}$$

then, the maximizing vector for problem (30) is given by

$$\begin{cases} x_i^* = \dfrac{\lambda_r - b_i}{2c_i} & (i = 1, 2, \ldots, r) \\ x_i^* = 0 & (i = r+1, r+2, \ldots, m) \; . \end{cases} \tag{33}$$

From (31) we get

$$\lambda_r = \frac{C + \sum_{i=1}^{r} \dfrac{b_i}{2c_i}}{\sum_{i=1}^{r} \dfrac{1}{2c_i}} \tag{34}$$

and from (32) - (34) it follows

$$C > \sum_{i=1}^{r} \frac{b_r - b_i}{2c_i} \quad . \tag{35}$$

Thus, if r is the greatest index i, i = 1, 2, ..., m such that (35) holds, then (33) gives the maximizing vector and λ_r is in (34).

b) Let us solve the problem [8]

$$\max \sum_{i=1}^{m} p_i (1 - e^{-k_i x_i}) \tag{36}$$

$$\begin{cases} \sum_{i=1}^{m} x_i \leqslant C \\ x_i \geqslant 0 \end{cases}$$

with $C > 0$ and $p_i > 0$.

As the components x_i^* of the maximizing vector, corresponding to $k_i < 0$, are zero, we can assume

$$k_i > 0 \qquad i = 1, 2, \ldots, m$$

and

$$p_1 k_1 \geq p_2 k_2 \geq \cdots \geq p_m k_m$$

in order to get (21).

Let r be the greatest index i, i = 1, 2, ..., m, such that the equation

$$\sum_{i=1}^{r} \frac{\log p_i k_i - \log \lambda}{k_i} = C \tag{37}$$

has a root λ_r, satisfying the inequality

$$p_r k_r > \lambda_r . \tag{38}$$

Then, the vector

$$\begin{cases} x_i^* = \dfrac{\log p_i k_i - \log \lambda_r}{k_i} & (i = 1, 2, 3, \ldots, r) \\ x_i^* = 0 & (i = r+1, r+2, \ldots, m) \end{cases} \tag{39}$$

solves our problem.

Indeed, it is easy to show that, from (37) - (38) - (40)

$$\log \lambda_r = \frac{\displaystyle\sum_{i=1}^{r} \frac{\log p_i k_i}{k_i} - C}{\displaystyle\sum_{i=1}^{r} \frac{1}{k_i}} , \tag{40}$$

we obtain

$$C > \sum_{i=1}^{r} \frac{\log \dfrac{p_i k_i}{p_r k_r}}{k_i} . \tag{41}$$

Thus, if r is the greatest index i, i = 1, 2, ..., m such that (41) holds, then (39) gives the maximizing vector for λ_r satisfying (40).

c) Finally, trying to solve the problem [9]

$$\text{Max} \sum_{i=1}^{m} \frac{p_i x_i}{x_i + k_i} \tag{42}$$

$$\begin{cases} \sum_{i=1}^{m} x_i \leq C \\ x_i \geq 0 \end{cases}$$

with $C > 0$, $p_i > 0$, $k_i > 0$, we can assume the sequence

$$\frac{p_1}{k_1} \geq \frac{p_2}{k_2} \geq \dots \geq \frac{p_m}{k_m}$$

in order to satisfy (21).

Under these hypotheses, if r is the greatest index i, i = 1, 2, ..., m, such that, for a root of the equation

$$\sum_{i=1}^{r} \left(\frac{k_i p_i}{\lambda}\right)^{\frac{1}{2}} - k_i = C \tag{43}$$

the following inequality holds

$$\frac{p_r}{k_r} > \lambda_r , \tag{44}$$

then, the point

$$\begin{cases} x_i^* = \left(\dfrac{k_i p_i}{\lambda_r}\right)^{\frac{1}{2}} - k_i & (i = 1, 2, \ldots, r) \\ x_i^* = 0 & (i = r+1, r+2, \ldots, m) \end{cases} \tag{45}$$

solves problem (42).

From (43) we get the relationship

$$\lambda_r = \left(\frac{\sum_{i=1}^{r} (k_i p_i)^{\frac{1}{2}}}{C + \sum_{i=1}^{r} k_i} \right)^2 , \tag{46}$$

and, by means of (44) and (46), we obtain the inequality

$$C > \frac{\sum_{i=1}^{r} (k_i p_i)^{\frac{1}{2}}}{\left(\dfrac{p_r}{k_r}\right)^{\frac{1}{2}}} - \sum_{i=1}^{r} k_i \tag{47}$$

Thus, if r is the greatest index i, i = 1, 2, ..., m such that (47) holds, then the maximizing vector is given by (45) with λ_r assigned by (46).

6. SOLUTION OF THE CLASSICAL ALLOCATION PROBLEM WITH ADDITIONAL CONSTRAINTS ON VARIABLES, $0 \leq x_i \leq \alpha_i$.

In this section, we deal with problem

$$\max \sum_{i=1}^{m} f_i(x_i) \tag{48}$$

$$\begin{cases} \sum_{i=1}^{m} x_i \leq C \\ 0 \leq x_i \leq \alpha_i \qquad (i = 1, 2, \ldots, m) \end{cases}$$

under the usual assumption that $f_i(x_i)$ are differentiable, strictly concave functions and, with no loss of generality, satisfy inequalities $f'_i(0) > 0$ and (21).

First of all, let us assume all the functions $f_i(x_i)$ increase in $[0, C]$. If the upper bounds α_i satisfy inequality $\sum_{i=1}^{m} \alpha_i < C$, the solution of (48)

$$x_i^* = \alpha_i \qquad (i = 1, 2, \ldots, m)$$

is trivial.

Alternatively, when we meet inequality $\sum_{i=1}^{m} \alpha_i > C$, the solution of problem (48) satisfies the constraint $\sum_{i=1}^{m} x_i \leq C$ in equational form and we are led to the equivalent problem

$$\max \sum_{i=1}^{m} f_i(x_i) \tag{49}$$

$$\begin{cases} \sum_{i=1}^{m} x_i = C \\ 0 \leq x_i \leq \alpha_i \qquad (i = 1, 2, \ldots, m) \end{cases}$$

For (49), the Kuhn - Tucker conditions are given by

$$\begin{cases} f_i'(x_i) \leqslant \lambda + \mu_i & (i = 1, 2, \ldots, m) , \\ x_i \geqslant 0 , & \\ x_i f_i'(x_i) = x_i(\lambda + \mu_i) & (i = 1, 2, \ldots, m) , \\ \sum_{i=1}^{m} x_i = C , & \\ x_i \leqslant \alpha_i & (i = 1, 2, \ldots, n) , \\ \mu_i \geqslant 0 & (i = 1, 2, \ldots, m) , \\ \mu_i(x_i - \alpha_i) = 0 & (i = 1, 2, \ldots, m) : \end{cases} \tag{50}$$

if there is no upper bound condition on variables in (49), system (50) becomes

$$\begin{cases} f_i'(x_i) \leqslant \lambda & (i = 1, 2, \ldots, m) , \\ x_i \geqslant 0 & (i = 1, 2, \ldots, m) , \\ x_i f_i'(x_i) = x_i \lambda & (i = 1, 2, \ldots, m) , \\ \sum_{i=1}^{m} x_i = C & \end{cases} \tag{51}$$

and we evaluate its solution point

$$(x_1^*, \ldots, x_m^*, \lambda^*)$$

by means of technique described in section 4.

For $x_i^* \leq \alpha_i$, $i = 1, 2, \ldots, m$, the m-dimensional vector $(x_1^*, \ldots, x_m^*)$ solves problem (49) and the (2m+1)-dimensional vector

$$(x_1^*, \ldots, x_m^*, \lambda^*, \mu_1^*, \ldots, \mu_m^*)$$

with $\mu_i^* = 0$ for $i = 1, 2, \ldots, m$, solves system (50).

For $x_j^* > \alpha_j$, $j \in I_1 \subset I = \{1, 2, \ldots, m\}$, we have to consider problem

$$\text{Max} \sum_{h \in I-I_1} f_h(x_h) \tag{52}$$

$$\begin{cases} \sum_{h \in I-I_1} x_h = C - \sum_{j \in I_1} \alpha_j \\ x_h \geq 0 \quad (h \in I-I_1) \end{cases}$$

for which the Kuhn - Tucker conditions are given by

$$\begin{cases} f_h'(x_h) \leq \lambda & (h \in I - I_1) , \\ x_h \geq 0 & (h \in I - I_1) , \\ x_h f_h'(x_h) = x_h \lambda & (h \in I - I_1) , \\ \sum_{h \in I-I_1} x_h = C - \sum_{h \in I_1} \alpha_j . \end{cases} \tag{53}$$

Let $(x_h^{**}, \lambda^{**} \mid h \in I - I_1)$ solve system (53), according to section 4.

If we find $x_h^{**} \leq \alpha_h$ for any $h \in I - I_1$, the m-dimensional vector

$$(\alpha_j,\ x_h^{**} \mid j \in I_1,\ h \in I - I_1)\ , \qquad (54)$$

solves problem (49). Indeed, let us assume

$$\begin{cases} \mu_j^{**} = f_j'(\alpha_j) - \lambda^{**} & \text{for any } j \in I_1 \\ \mu_h^{**} = 0 & \text{for any } h \in I - I_1 \end{cases}$$

As $f_j'(x_j)$ is a decreasing function and $x_j^* > \alpha_j$, the following inequalities hold

$$\mu_j^{**} = f_j'(\alpha_j) - \lambda^{**} > f_j'(x_j^*) - \lambda^{**} = \lambda^* - \lambda^{**}\ ,$$

the last of which follows from

$$x_i f_i'(x_i) = x_i \lambda \qquad \text{for } i = 1, 2, \ldots, m\ .$$

For the particular assumption made on $f_i(x_i)$, we can notice that λ^* and λ^{**} are positive because of (51) and (53).

We are now going to prove the relationship $\lambda^* \geqslant \lambda^{**}$.

From inequality $C - \sum_{j \in I_1} \alpha_j > C - \sum_{j \in I_1} x_j^*$, we get

$$x_h^{**} \geqslant x_h^* \qquad \text{for any } h \in I - I_1\ , \qquad (55)$$

one of which, at least, is strict inequality.

If there esists at least one index $h \in I - I_1$ such to have $x_h^* > 0$, then, $f_h'(x_h)$ being a decreasing function, from the third relation in (51) and (53) respectively, we obtain $\lambda^* \geqslant \lambda^{**}$.

On the other hand, if we have $x_h^* = 0$ for any $h \in I - I_1$, at least the first component x_h^{**}, $x_{h_1}^{**}$, is positive, under the assumption that $f_h(x_h)$ are ordered according to (21).

Finally, by using relationships $f'_{h_1}(0) \leq \lambda^*$ and $f'_{h_1}(x_{h_1}^{**}) = \lambda^{**}$, $f'_{h_1}(x_{h_1})$ being a decreasing function, we have $\lambda^* > \lambda^{**}$.

We can now conclude that inequality $\mu_j^{**} > 0$ is true and the m-dimensional vector $(\alpha_j, x_h^{**} \mid j \in I_1, h \in I - I_1)$ solves system (49).

Let us now assume $x_h^{**} > \alpha_h$ at least for an index $h \in I - I_1$: in this case we iterate the procedure described above and reach the solution point for (49) in m-1 steps, at most.

In order to conclude the description of our algorithm, we have to assume that not all the functions $f_i(x_i)$ are increasing in $[0, C]$: in this case we solve the equivalent problem

$$\max \sum_{i=1}^{m} f_i(x_i) \tag{56}$$

$$\begin{cases} \sum_{i=1}^{m} x_i \leq C \\ 0 \leq x_i \leq \gamma_i \qquad (i = 1, 2, \dots, m) \end{cases}$$

where $\gamma_i = \min(\alpha_i, \beta_i)$ and β_i is either the root of equation $f'_i(x_i) = 0$ on the open - closed interval $(0, C]$, if it exists, or $\beta_i = +\infty$, if such a root does not exist.

On interval $[0, \gamma_i]$, any function $f_i(x_i)$ is increasing and we can use the solution technique described above.

7. A MULTISTAGE PROCEDURE.

We can solve problem (13) by applying a dynamic programming algorithm, i.e. by a multistage procedure that, at each step, solves the problem

$$F_i(z) \overset{\text{def}}{===} \underset{0 \leq x_i \leq z}{\text{Max}} \left[F_{i-1}(z - x_i) + f_i(x_i) \right] \quad (i = 2, 3, \ldots, m)$$

with

$$F_1(z) = f_1(z) \quad \text{and} \quad 0 \leq z \leq C .$$

Lemmas proved in section 3, allow us to reduce the amount of computations. Indeed, at each stage, we can decide whether to examine the following steps.

First of all, let us set

$$G_i(x_i) \overset{\text{def}}{===} F_{i-1}(z - x_i) + f_i(x_i) \qquad 0 \leq x_i \leq z . \tag{57}$$

If $\hat{x}_i$ maximizes $G_i(x_i)$ over $[0, z]$ and $\hat{x}_j$ $(j = 1, 2, \ldots, i-1)$ solves the problem

$$F_{i-1}(z - \hat{x}_i) = \max \sum_{j=1}^{i-1} f_j(x_j)$$

$$\begin{cases} \sum_{j=1}^{i-1} x_j = z - \hat{x}_i \\ x_j \geq 0 \end{cases} \qquad (j = 1, 2, \ldots, i-1)$$

then $(\hat{x}_1, \hat{x}_2, \ldots, \hat{x}_{i-1}, \hat{x}_i)$ satisfies (22) for $i = m$, by keeping (21).

We are now going to prove two interesting theorems.

<u>Theorem 1.</u>

Let the functions $f_i(x_i)$ *satisfy* (21).

Then for only one number α*, at most, with* $0 \leqslant \alpha < z$*, we get*

$$G_i'(x_i) > 0 \qquad \textit{for} \quad 0 \leqslant x_i < \alpha \ \textit{if} \ \alpha > 0$$

$$G_i'(x_i) = 0 \qquad \textit{for} \quad x_i = \alpha$$

$$G_i'(x_i) < 0 \qquad \textit{for either} \ \alpha < x_i \leqslant z \ \textit{or} \ 0 \leqslant x_i \leqslant z, \ \textit{if} \ \alpha \ \textit{does not exist.}$$

Proof.

First of all, as $F_{i-1}(z - x_i)$ is a strictly concave function with respect to x_i [6], $G_i(x_i)$ satisfies the same property and has only one maximizing point over $[0, z]$.

Moreover, the inequality $G_i'(z) \geqslant 0$ contradicts (22): under this condition $G_i(x_i)$ reaches its maximum value for $\hat{x}_i = z$ and $\hat{x}_j = 0$, $j = 1, 2, \ldots, i-1$, and this cannot hold for Lemma 2.

Thus we are forced to assume

$$G_i'(z) < 0 \ . \tag{58}$$

If we assume $G_i'(z) \geqslant 0$, then $G_i(x_i)$, which is strictly concave, increases in the whole interval $[0, z]$ and reaches its maximum value at $\hat{x}_i = z$: that contradicts (21). Thus it cannot increase over the whole interval $[0, z]$ and either it decreases in $[0, z]$ or it increases in a closed - open interval $[0, \alpha)$, reaches its maximum value at α and decreases in the open-closed interval $(\alpha, z]$.

It is easy to see that it is $\alpha \neq z$, according to (58).

Theorem 2.

Under condition (21), *let* $\hat{x}_i = 0$ *maximize* $G_i(x_i)$ *in* $[0, z]$ *and* $\hat{x}_{i+1}$

maximize $G_{i+1}(x_{i+1})$ *in* $[0, z]$, *too. Then* $\hat{x}_{i+1} = 0$.

Proof.

Indeed, let $x_i^{(i+1)}$ maximize $G_i(x_i)$ in $\left[0, z - \hat{x}_{i+1}\right]$: from theorem 1 the condition $\hat{x}_i = 0$ implies that $G_i(x_i)$ decreases in $[0, z]$ and in $\left[0, z - \hat{x}_{i+1}\right]$, too.

Thus we get $x_i^{(i+1)} = 0$ and, for lemma 2, $\hat{x}_{i+1} = 0$.

8. SOLUTION OF PROBLEM (13) BY MEANS OF A DYNAMIC PROGRAMMING ALGORITHM.

Let us solve problem (13), still under assumption (21).

At the first step, we define

$$F_1(z) \overset{\text{def}}{=\!=\!=} \max_{x_1=z} f_1(x_1) = f_1(z) \qquad 0 \leqslant z \leqslant C \ ;$$

at the second step, we solve the problem

$$F_2(z) \overset{\text{def}}{=\!=\!=} \max_{0 \leqslant x_2 \leqslant z} \{F_1(z - x_2) + f_2(x_2)\} \qquad 0 \leqslant z \leqslant C \ ,$$

i.e. we maximize the function

$$G_2(x_2) = F_1(z - x_2) + f_2(x_2) \ ,$$

over the interval $[0, z]$.

From the equality

$$G'(x_2) = -F_1'(z - x_2) + f_2'(x_2) \ ,$$

we get

$$G_2'(0) \gtreqless 0 , \tag{59}$$

if and only if (°)

$$z \gtreqless F_1'^{-1}(f_2'(0)) , \tag{60}$$

and, for theorem 1, $G_2(x_2)$ reaches its maximum value over $[0, z]$ either at the point

$$\hat{x}_2 = 0 \quad \text{if} \quad z \leqslant F_1'^{-1}(f_2'(0)) ,$$

or at the point

$$\hat{x}_2 = \alpha_2 \quad \text{if} \quad z > F_1'^{-1}(f_2'(0)) ,$$

α_2 being the only root of the equation

$$F_1'(z - x_2) = f_2'(x_2) . \tag{61}$$

Thus the function $\sum_{i=1}^{2} f_i(x_i)$ gets its maximum value under the constraints $\sum_{i=1}^{2} x_i = z$ and $x_i \geqslant 0$ either at the point

$$\begin{cases} \hat{x}_2 = 0 \\ \hat{x}_1 = z \end{cases} \quad \text{if } z \leqslant F_1'^{-1}(f_2'(0)) ,$$

(°) As $F_i(z)$ $(i = 1, 2, \ldots, m)$ is a strictly concave function, then $F_i'(z)$ decreases and can be inverted.

or at the point

$$\begin{cases} \hat{x}_2 = \alpha_2 \\ \hat{x}_1 = z - \hat{x}_2 \end{cases} \qquad \text{if } z > F_1'^{-1}(f_2'(0)) \ .$$

It is worthwile to remark that, if it holds

$$C \leqslant F_1'^{-1}(f_2'(0)) \ , \tag{62}$$

the point which maximizes $G_2(x_2)$ over $[0, C]$ is $x_2^* = 0$ and theorem 2 allows us not to examine the following stages, as the optimal policy is given by

$$\begin{cases} x_1^* = C \\ x_j^* = 0 \end{cases} \qquad (j = 2, 3, \ldots, m)$$

and the corresponding maximum value of the objective function is

$$F_m(C) = F_1(C) + \sum_{i=2}^{m} f_i(0) = f_1(C) + \sum_{j=2}^{m} f_j(0) \ .$$

On the other hand, when (62) does not hold, we need solve the next problem

$$F_3(z) = \max_{0 \leqslant x_3 \leqslant z} \left\{ F_2(z - x_3) + f_3(x_3) \right\} \ .$$

For our purpose, let us set

$$G_3(x_3) = F_2(z - x_3) + f_3(x_3) :$$

from the relationship

$$G_3'(x_3) = -F_2'(z - x_3) + f_3'(x_3) ,$$

we get

$$G_3'(0) \gtreqless 0 \tag{63}$$

if and only if

$$z \gtreqless F_2'^{-1}(f_3'(0)) . \tag{64}$$

For theorem 1 and relationships (63)-(64), $G_3(x_3)$ is maximum over $[0, z]$ either at the point

$$\hat{x}_3 = 0 \quad \text{if } z \leq F_2'^{-1}(f_3'(0)) ,$$

or at the point

$$\hat{x}_3 = \alpha_3 \quad \text{if } z > F_2'^{-1}(f_3'(0)) ,$$

α_3 being the only root of the equation

$$F_2'(z - x_3) = f_3'(x_3) . \tag{65}$$

Thus, if $z \leq F_2'^{-1}(f_3'(0))$, the point, which maximizes $\sum_{i=1}^{3} f_i(x_i)$ under the constraints $\sum_{i=1}^{3} x_i = z$, $x_i \geq 0$, is given by

$$\begin{cases} \hat{x}_3 = 0 \\ \hat{x}_2 = \alpha_2 \\ \hat{x}_1 = z - \hat{x}_2 \end{cases} ,$$

where α_2 solves equation (61); alternatively, if $z > F_2'^{-1}(f_3'(0))$, the solution is

$$\begin{cases} \hat{x}_3 = \alpha_3 \\ \hat{x}_2 = \alpha_2 \\ \hat{x}_1 = z - \hat{x}_2 - \hat{x}_3 \end{cases} ,$$

where α_3 solves (65) and α_2 is a root of the equation

$$F_1'(z - \hat{x}_3 - x_2) = f_2'(x_2) .$$

In this case too, if it is

$$C \leq F_2'^{-1}(f_3'(0)) \quad ,$$

we need not examine all the following steps of the algorithm, because $x_3^* = 0$ maximizes $G_3(x_3)$ over $[0, C]$ and the optimal policy is

$$\begin{cases} x_j^* = 0 & (j = 3, 4, \ldots, m) \\ x_2^* = \alpha_2 \\ x_1^* = C - x_2^* \end{cases}$$

where α_2 is the only root of (65) with $z = C$ and the maximum value of the objective function is equal to

$$F_m(C) = F_2(C) + \sum_{j=3}^{m} f_j(0) \quad .$$

At the i-th stage, under the assumption (°)

$$C > F'^{-1}_{i-2}(f'_{i-1}(0)) \ , \tag{66}$$

we solve the problem

$$F_i(z) \overset{\text{def}}{=\!=\!=} \max_{0 \leqslant x_i \leqslant z} \left\{ F_{i-1}(z - x_i) + f_i(x_i) \right\} \quad .$$

Let us set

$$G_i(x_i) = F_{i-1}(z - x_i) + f_i(x_i) \ :$$

then, by using the equality

$$G'_i(x_i) = -F'_{i-1}(z - x_i) + f'_i(x_i) \ ,$$

it follows

$$G'_i(0) \gtreqless 0 \ , \tag{67}$$

(°) If (57) does not hold, the maximizing point of (13) is given by the one of the (i-2)-th stage, taken at $z = C$ and completed by $x_j = 0$ for $j = i-1, i, \ldots, m$.

if and only if

$$z \gtreqless F'^{-1}_{i-1}\,(f'_i(0)) \,. \tag{68}$$

Theorem 1 and relationships (67)-(68) allow us to infer that the maximizing point of $G_i(x_i)$ over $[0, z]$ is either

$$\hat{x}_i = 0 \qquad \text{if } z \leqslant F'^{-1}_{i-1}\,(f'_i(0)) \quad ,$$

or

$$\hat{x}_i = \alpha_i \qquad \text{if } z > F'^{-1}_{i-1}\,(f'_i(0)) \quad ,$$

where α_i is the only root of the equation

$$F'_i(z - x_i) = f'_i(x_i) \,.$$

Thus, the solution point for the problem of maximizing the function $\sum_{j=1}^{i} f_j(x_j)$ under the constraints $\sum_{j=1}^{i} x_j = z$ and $x_j \geqslant 0$, is given by

$$\begin{cases} \hat{x}_i = 0 \\ \hat{x}_j = \alpha_j \qquad\qquad (j = i-1,\ i-2,\ \dots,\ 2) \ , \\ \hat{x}_1 = z - \sum_{j=2}^{i-1} \hat{x}_j \end{cases}$$

if $z \leqslant F'^{-1}_{i-1}\,(f'_i(0))$, α_j being the root of the equation

$$F'_{j-1}(z - \sum_{h=j}^{i-2} \hat{x}_h - x_j) = f'_j(x_j) \qquad (j = i-1,\ i-2,\ \dots,\ 2) \tag{69}$$

with $\sum_{h=j}^{i-2} \hat{x}_h = 0$ for $j = i-1$. On the contrary, when we find $z > F'^{-1}_{i-1}(f'_i(0))$, the maximizing point is given by

$$\begin{cases} \hat{x}_j = \alpha_j & (j = i, i-1, \dots, 2) \\ \hat{x}_1 = z - \sum_{j=2}^{i} \hat{x}_j \end{cases}$$

α_j being the root of the equation

$$F'_{j-1}(z - \sum_{h=j}^{i-1} \hat{x}_h - x_j) = f'_j(x_j) \qquad (j = i, i-1, \dots, 2)$$

where $\sum_{h=j}^{i-1} \hat{x}_h = 0$ for $j = 1$.

In this general step too, we can see that the inequality

$$C \leqslant F'^{-1}_{i-1} (f'_i(0))$$

allows us not to consider the following steps, because $G_i(x_i)$ is maximum at $x^*_i = 0$ over $[0, C]$ and, consequently, problem (13) is solved by

$$\begin{cases} x^*_j = 0 & (j = i, i+1, \dots, m) \\ x^*_j = \alpha_j & (j = i-1, i-2, \dots, 2), \\ x^*_1 = C - \sum_{j=2}^{i-1} x^*_j \end{cases}$$

α_j being the root of (69) with $z = C$, and the maximum value for the objective function is

$$F_m(C) = F_{i-1}(C) + \sum_{j=i}^{m} f_j(0) .$$

Following the method described above, we can get the solution of (13) in m steps at most.

Thus, if r is the greatest index i = 1, 2, ..., m, such that the following inequality holds

$$C > F'^{-1}_{r-1} (f'_r(0)) ,$$

the only solution point for (13) is given by

$$\begin{cases} x_i^* = 0 & (i = r+1, r+2, \dots, m) \\ x_i^* = \alpha_i & (i = 2, 3, \dots, r) \quad , \\ x_i^* = C - \sum_{i=2}^{r} x_i^* \end{cases}$$

α_i being the only root of the equation

$$F'_{i-1} (C - \sum_{j=r+1}^{r} x_j^* - x_i) = f'_i(x_i) \qquad (i = r, r-1, \dots, 2) ,$$

with $\sum_{j=i+1}^{r} x_j^* = 0$ for i = r.

REFERENCES

1. Bellman, R., *Dynamic Programming*, Princeton University Press, 1967.

2. Volpato, M., *Studi e modelli di ricerca operativa*, U.T.E.T., Torino, 1971, chap. 2.

3. Lisei, G., *Su un particolare problema di ricerca operativa risolto con il lemma di J. Willard Gibbs*, Department of Mathematics, University of Genoa, 1972.

4. Volpato, M., *Studi e modelli di ricerca operativa*, U.T.E.T., Torino, 1971, chap. 7, pg. 1052.

5. Giannessi, F., Alcune considerazioni sulla risoluzione di classici problemi di riassicurazione, in Volpato [2], chap. 3.

6. Giannessi, F., Sulla risoluzione col metodo della programmazione dinamica di un problema di estremo concernente la scadenza media, in Volpato [2], chap. 3.

7. Cucconi, O., Sopra un particolare problema (non lineare) di distribuzione, in Volpato [2], chap. 3.

8. Wilkinson, C. and Gupta, S.K., Allocation Promotional Effort to Competing Activities: A Dynamic Approach, in *Proceedings of the Vth International Conference on Operational Research, Venice 1969*, J. Lawrence ed., Tavistock Publications, London, 1970.

9. Castellani, G., Su un particolare problema di teoria della ricerca, *Rendiconti del Comitato per il potenziamento in Venezia degli studi economici*, CEDAM, Padova, 1971.

Erratum of W. Stadler

Natural Structural Shapes, p. 95.

The statement beginning with " Thus, ... " preceding the inequality (3.49) is incorrect. As a matter of fact one can prove the following about the two designs (see [28] for the details):

(1) For a given mass

$\bar{\delta} \leqq \delta^*, \quad g_2(u^*(\cdot)) \leqq g_2(\bar{u}(\cdot)), \quad \tau^*_{min} \leqq \bar{\tau} \leqq \tau^*_{max}$.

(2) For a given maximum deformation

$g_1(\bar{u}(\cdot)) \leqq g_1(u^*(\cdot)), \quad g_2(u^*(\cdot)) \leqq g_2(u(\cdot)), \quad \tau^*_{min} \leqq \bar{\tau} \leqq \tau^*_{max}$.

(3) For a given maximum stress

$g_1(\bar{u}(\cdot)) \leqq g_1(u^*(\cdot)), \quad g_2(u^*(\cdot)) \leqq g_2(\bar{u}(\cdot)), \quad \delta^* \leqq \bar{\delta}$.

(4) The natural structure is the minimum weight structure for a given stored energy.

(5) When k_1 and k_2 are considered as small parameters in the solution for the natural shape, then the constant stress structure is a first-order approximation to the natural structure in terms of these parameters.

(6) In the limit as g_1 tends to infinity, $\bar{\delta}$ and $g_2(\bar{u}(\cdot))$ tend to zero. However, δ^* tends to $\frac{1}{2}\, k_1 k_2$ and $g_2(u^*(\cdot))$ tends to $\frac{1}{2}\, k_2 \omega$, a more reasonable result, since it clearly makes no sense to have a loaded bar whose deformation may be made arbitrarily small by loading it with more mass.

Printed in the United States
By Bookmasters